The Oil and Gas Service Industry in Asia

Palgrave Macmillan Asian Business Series Centre for the Study of Emerging Market Series

Series Editor: **Harukiyo Hasegawa** is Professor at Doshisha Business School, Kyoto, Japan, and Honourable Research Fellow at the University of Sheffield's School of East Asian Studies, where he was formerly Director of the Centre for Japanese Studies.

The Palgrave Macmillan Asian Business Series seeks to publish theoretical and empirical studies that contribute forward-looking social perspectives on the study of management issues not just in Asia, but by implication elsewhere. The series specifically aims at the development of new frontiers in the scope, themes and methods of business and management studies in Asia, a region which is seen as key to studies of modern management, organisation, strategies, human resources and technologies.

The series invites practitioners, policy-makers and academic researchers to join us at the cutting edge of constructive perspectives on Asian management, seeking to contribute towards the development of civil societies in Asia and further afield.

Titles include:
Glenn D. Hook and Harukiyo Hasegawa (*editors*)
JAPANESE RESPONSES TO GLOBALIZATION IN THE 21st CENTURY
Politics, Security, Economics and Business

Diane Rosemary Sharpe and Harukiyo Hasegawa (*editors*)
NEW HORIZONS IN ASIAN MANAGEMENT
Emerging Issues and Critical Perspectives

Sten Söderman (*editor*)
EMERGING MULTIPLICITY
Integration and Responsiveness in Asian Business Development

Oliver H.M. Yau and Raymond P.M. Chow (*editors*)
HARMONY VERSUS CONFLICT IN ASIAN BUSINESS
Managing in a Turbulent Era

Tan Yi
THE OIL AND GAS SERVICE INDUSTRY IN ASIA
A Comparison of Business Strategies

Palgrave Macmillan Asian Business Series
Series Standing Order ISBN 978-1-4039-9841-5

You can receive future titles in this series as they are published by placing a standing order. Please contact your bookseller or, in case of difficulty, write to us at the address below with your name and address, the title of the series and the ISBN quoted above.

Customer Services Department, Macmillan Distribution Ltd, Houndmills, Basingstoke, Hampshire RG21 6XS, England

---

The Oil and Gas Service Industry in Asia

A Comparison of Business Strategies

Tan Yi
Senior Market Analyst, ODS-Petrodata

First published 2010 by
PALGRAVE MACMILLAN

Palgrave Macmillan in the UK is an imprint of Macmillan Publishers Limited, registered in England, company number 785998, of Houndmills, Basingstoke, Hampshire RG21 6XS.

Palgrave Macmillan in the US is a division of St Martin's Press LLC, 175 Fifth Avenue, New York, NY 10010.

Palgrave Macmillan is the global academic imprint of the above companies and has companies and representatives throughout the world.

Palgrave® and Macmillan® are registered trademarks in the United States, the United Kingdom, Europe and other countries.

ISBN 978–0–230–23559–5   hardback

This book is printed on paper suitable for recycling and made from fully managed and sustained forest sources. Logging, pulping and manufacturing processes are expected to conform to the environmental regulations of the country of origin.

A catalogue record for this book is available from the British Library.

Library of Congress Cataloging-in-Publication Data

Yi, Tan, 1966–
   The oil and gas service industry in Asia : a comparison of business strategies / Tan Yi.
       p. cm.
     Summary: "This book investigates the business strategies chosen by oil and gas service companies operating in China, Singapore and Malaysia. It provides an analytical view of the reliability of strategic theoretical frameworks based on Western business practice but applied in a non-Western business environment like Asia"—Provided by publisher.
     Includes bibliographical references and index.
     ISBN 978–0–230–23559–5   (hardback)
  1. Petroleum industry and trade—China. 2. Petroleum industry and trade—Singapore. 3. Petroleum industry and trade—Malaysia. I. Title.

HD9576.C52Y5 2010
338.4'7622338095—dc22   2009046800

10   9   8   7   6   5   4   3   2   1
19   18   17   16   15   14   13   12   11   10

Printed and bound in Great Britain by
CPI Antony Rowe, Chippenham and Eastbourne

Contents

List of Figures

List of Tables

Acknowledgements

I would like to take this opportunity to express my sincere thanks to Douglas Gourlay, Ted Mason and Alex Wilson for their supervision and input on this PhD research project. Special thanks go to Jeremy Cresswell for his editorial assistance and general help.

I am most grateful that Simmons & Company International, ASCO plc, and the Robert Gordon University awarded me a studentship for research studies.

Over 200 senior executives of energy service organisations from Asia, Europe and North America took part in the empirical research and 24 Chinese executives participated in the pilot fieldwork in China. Without their contribution, successful completion of the research project could never have been achieved. The need for anonymity prevents me from listing them individually.

I would also like to thank Palgrave Macmillan for supporting the publication of this book and Professor Harukiyo Hasegawa for his encouragement and insights on my work.

Tan Yi
Aberdeen, Scotland

List of Abbreviations

bcf	billion cubic feet
bcm	billion cubic metres
BOE	barrels oil equivalent
CAGR	compound annual growth rate
Capex	capital expenditure
CCP	Chinese Communist Party
COHC	CITIC Offshore Helicopter Company
COOEC	China Offshore Oil Engineering Company Limited
COSL	China Oilfield Services Limited
CPOE	China Petroleum Offshore Engineering Company
CPP	China Petroleum Pipeline Bureau
CSIC	China Shipbuilding Industry Corporation
CSSC	China State Shipbuilding Corporation
E&P	exploration and production
EIA	Energy Information Administration
EMS	emerging medical service
EPC	engineering, procurement and construction
EPCM	engineering, procurement and construction management
EPIC	engineering, procurement, installation and construction
ESP	environment, strategy and performance
FEED	front-end engineering and design
FPSO	floating production, storage and offloading
FPU	floating production unit
IMR	inspection, maintenance and repair
IOC	international oil company

JDA	joint development area
JV	joint venture
MISC	Malaysia International Shipbuilding Corporation
MMHE	Malaysia Marine and Heavy Engineering
NOC	national oil company
OPEC	Organisation of Petroleum Exporting Countries
PRC	People's Republic of China
PSA	production sharing agreement
PSC	production sharing contract
SAR	search and rescue
SETC	State Economic and Trade Commission
SME	small to medium-sized enterprise
SPSS	statistical package for the social sciences
SOE	state-owned enterprise
SPC	Singapore Petroleum Corporation
TD	target depth
WTI	West Texas Intermediate

1
Introduction

The importance of understanding the business environment and adapting strategy to achieve a high standard of performance is widely acknowledged in the strategic management literature. However, business strategy theory has largely been founded on research into organisations that operate in a Western business environment. Consequently, it may have little practical application in Asia, which is a major region for oil and gas activities.

China has become East Asia's leading economic powerhouse and is also the regional leader in terms of oil and gas exploration and production (E&P). For several decades, the country has exhibited enviable growth. Since the growth of the economy ultimately boils down to the success of Chinese firms, some believe that firms in this region must have employed strategies conducive to such performance. Previous research (Peng et al., 2001) shows that firms in China display distinctive business strategies that dazzle and bewilder the outside world. Nonetheless, despite real strengths, some state-owned Chinese companies may fail to perform well. Their business competences still need to be improved through reassessing and resetting their strategies. In many indigenous Chinese companies, there are no systematic or articulated strategies and no formal strategic planning process.

It is this that stimulated the idea of conducting investigations into what business strategies are applied by energy service firms in East Asia. In order to develop typologies grounded in organisational and strategic management theory, this empirical research seeks to study executives' perceptions on the business environment and their strategic orientations in China, and to a lesser extent, Singapore and Malaysia.

1.1   Research context

1.1.1   Defining the oil and gas service industry

In this book, the oil and gas service industry is defined as a group of entities engaged in the business of providing products or services to an oil and gas company, another oil service company, or to another energy sector (Simmons and Company International, 1999). It comprises a group of professional companies that often engage in highly specialised supply and service work in the petroleum industry. Integrated contractors offer a portfolio of interlinked services and products, all geared towards client petroleum companies or other oil and gas service companies (Phillips, 1998). Some service organisations such as Halliburton, a technologically focused oilfield service conglomerate, cover the complete value chain from exploration via field development to production.

According to Phillips (1998), the service industry comprises a number of subsets serving various market segments. Within each of the subsectors, service firms usually have their own specialised (focused) area and, traditionally, they rarely become involved with other activities though vertical integration of services is becoming increasingly common. They provide related and highly specialised services such as cementing, casing into a well, pumping acid into a well, or drilling a well. Frequently, contractors and service suppliers design, manufacture and own equipment to enable them to perform their specialist service. They generally have highly trained and experienced personnel to carry out the work. A service company in one subsector might have no knowledge or interest in any other service activities.

Phillips also suggests that the service industries mainly comprise contracting and construction companies, service companies, and vendors. They may include drilling contractors, well equipment specialists, component supply, fabrication and engineering, and subsea contractors. Contracting and construction companies carry out the design, fabrication and construction of infrastructure such as offshore platforms or production vessels (e.g. Aker Solutions and Sembcorp Marine), pipelines (e.g. Acergy and Baosteel), process equipment (e.g. BJ Services), and associated electrical and mechanical services used in the oil and gas industry (e.g. NATCO Group).

Service companies carry out a wide range of specialist services. In this study, the service businesses include:

- drilling of wells by drilling contractors;
- services associated with drilling and evaluating wells, including casing, cementing, logging, and testing;
- specialist technical services in support of platform operations, including instrumentation, helicopter and ship operations as well as completion, fishing and workover solutions;
- general services such as catering and cleaning, deck labour and crews.

In each of these market segments, there tend to be a number of dominant players.

Service companies serve the needs of the wider oil and gas industry and, in some instances, other sectors of the energy industry, such as nuclear, coal, power generation, renewables and petrochemicals. Many of them are also involved in business outside the energy industry. Rolls-Royce, for example, along its product lines, is divided into four business divisions serving different industrial sectors: energy, marine, civil aerospace and defence aerospace.

Vendors only produce equipment. Some of the equipment may be used in other industries like diving, electronics, and cabling and communications.

1.1.2 Oil and gas service sector: East Asia overview

East Asia[1] is now the second most important oil and gas consumption region after North America. China alone is a massive marketplace for the energy service industry. Wolf (2003) believes that China will potentially dwarf the US and the European Union to become the third global economy. Although coal remains China's dominant fuel, Beijing recognises that this is a major polluter and is therefore taking steps to deal with the problem by seeking to increase significantly the production and use of oil and gas, mostly as feedstock to raise electricity capacity. The potential market opportunities there have been becoming increasingly important to the oil and gas service sector.

The Chinese E&P industry has been generated by the ministries and state-owned enterprises (SOEs). This makes the Chinese oil and gas industry substantially though not entirely different from the

Western model. The most striking difference is that the Chinese national oil companies (NOCs) have almost everything integrated into them whereas Western international oil companies (IOCs) operate with a focus only on the oil and gas activities. Oil operators such as the China National Petroleum Corporation (CNPC) have their own research institutions, universities and service subsidiaries. Such service companies are not gathered into segment groupings, as they are in the West. In the Western model, service companies are mostly independent from oil operators though some such as Shell and Stateoil do spin off new technology businesses through joint ventures/joint industry projects.

In China, the 60-million strong Chinese Communist Party (CCP) leads the country. China's politics affect its business environment as one would expect. Within the energy sector, all industry regulation is government controlled (Scottish Enterprise, 2002). The Chinese government, for instance, has implemented a more restrictive safety and environmental protection policy offshore. All Chinese offshore service support companies must obtain the national standard certificate – Offshore Operations Permission. The certificate is required to be renewed annually. However, foreign service companies do not need to go through this procedure as they would be examined on an individual project basis by operators and the associated assessment reports would be submitted to the relevant government department (Tan, 2000). Nonetheless, it will almost certainly be necessary to have contact with government bodies in the early stages of approaching the Chinese market.

The island state of Singapore has no oil and gas reserves nor has there been any offshore activity. Like Japan, this country is a substantial importer of oil and gas (DTI, 1996). Nevertheless, it is unquestionably Asia's leading petroleum industry centre: 'It is the world's largest single bunkering port, the third largest refining centre after Houston and Rotterdam ... and the major oil and futures trading market for Asia' (The USA Singapore Embassy, 1997). Indeed, the country has become one of the three major global refining centres: US Gulf Coast (USGC), North West Europe (NWE) (Rotterdam) and Singapore (BP, 2002).

Although refining and petrochemical production are the backbone, bunkering, oil trading, and oil rig and floating production unit (FPU) manufacturing are important components of the industry.

Singapore is also regarded as an important logistics support base from which international service firms deliver their services to the region, mainly Southeast Asian E&P countries (Mackay and Adam, 1998; DTI, 1996; Abraham, 1999).

Despite the increasing competition from others in the region in refining and petrochemical production, Singapore remains the preferred regional hub for numerous international service and supply companies. The advantages of Singapore as a petroleum industry centre are: strong government support; good geographic location to serve Asia; well-established infrastructure such as highly efficient port services; economic, financial and political stability; and advanced information technology (Lucas, 2000; Mackay and Adam, 1998; BBC, 1994).

Malaysia is the most significant East Asian country in terms of offshore oil and gas activity and is an essential producer in the region (Mackay and Adam, 1998). Domestic oil production is offshore and primarily near Peninsular Malaysia. Activity in the states of Sarawak and Sabah off Kalimantan Island and in the joint development area (JDA) between Malaysia's Jerneh and Lawit fields and Thailand's Bongkot field is also substantial (DTI, 1996; EIA, 2001). One of the most active areas in Malaysia for gas exploration and development is the Malaysia–Thailand JDA, located in the lower part of the Gulf of Thailand and governed by the Malaysia–Thailand Joint Authority (MTJA) (EIA, 2001; BP, 2001).

Malaysia's deepwater reserves have doubled over the past five years. The most recent discovery offshore Sabah is of significant importance to future offshore production activities in the country. The substantial capital expenditure necessary for financing various discoveries has created new service opportunities such as subsea pipelines, oil and gas receiving terminals, and gas transmission pipelines (PETRONAS, 2009).

Whilst Malaysia is developing its own oil and gas expertise, by developing partnerships with foreign companies, it is also now looking to play a bigger part regionally. Many Malaysian companies are now developing their domestic base and seeking to become involved in playing an important role in the region of Southeast Asia and beyond.

In short, China, Singapore and Malaysia are well presented in terms of domestic-based manufacturing plants and service organisations, professional high-tech service companies from North America and Europe, and other foreign companies. All three countries are considered politically stable and in favour of moneymaking enterprise. As a result, they were considered fertile areas for investigation and thereby selected for the empirical work of this study.

Other countries in the region are much less significant oil and gas-related players and/or tend to be less financially and politically stable. The following summarises the service industry in each of these countries with brief explanations on why they were excluded from this study.

Like Malaysia, Indonesia has since the 1960s developed its own infrastructure and adopted policies supporting the development of domestic established companies. In spite of the fact that the country is very active as a substantial oil and gas producer, Indonesia is a politically and economically unstable country and has financial problems. Legal protection for enterprises is weak and the business environment has become more hostile for foreign investors (Zaobao, 2002). It was considered that Indonesia was a less appropriate selection than Malaysia and therefore excluded from this study.

Japan is one of the leading industrial countries in the world but its domestic energy sector is quite small compared with other East Asian countries. Although it has considerable manufacturing, engineering, software and wider technological capabilities to support the development of the oil and gas service business worldwide, like Singapore, Japan depends heavily on imported oil and gas. Since there is a language barrier with respect to communications, access to information and data collection, Japan was excluded from this research.

The Philippines are among the smaller oil and gas producers of East Asia. Between 1996 and 2000, only a handful or so of offshore exploration wells were drilling offshore. It also has financial problems. For these reasons, the Philippines were excluded from this study.

South Korea is one of the most prosperous and fastest-growing countries in the Asia-Pacific region but has no indigenous oil or gas production. The Korea Petroleum Development Corporation (PEDCO) has drilled a few offshore wells with no real success. Taiwan has a very small offshore oil and gas industry and little offshore activity is expected. Thus, neither of these two regions was considered.

Thailand is primarily a gas-producing country with a small level of oil production. Despite the E&P industry in this gas-rich country outperforming the overall economy in 1998 (Abraham, 1999), it also was not considered due to its financial problems and a low level of offshore exploration activity.

Vietnam is a significant country in the region in terms of offshore oil production and there is growing evidence of significant offshore gas reserves. It is a relatively smaller player compared to China and Malaysia and the country has experienced enduring political unrest. For the reasons applied elsewhere, Vietnam was not selected for the study.

1.1.3 Industry selection

The oil and gas industry is very broad and comprises three distinctive sectors:

- upstream covering exploration and production;
- midstream covering oil and gas transportation pipelines and tankers;
- and downstream covering refining, marketing and sales.

In this book, the market segments for oil and gas services are split in four: upstream, midstream, downstream and service sectors. Service companies frequently confine their activities variously to upstream, midstream or downstream, though there are plenty of major and mid-range firms that embrace at least two of these – upstream and midstream for example. It is a complex and dynamic network of facilities and organisations often with different and conflicting objectives.

According to Christopher (1998), 'the supply chain is the network of organisations that are involved, through upstream and downstream linkages, in the different processes and activities that produce value in the form of products and services in the hands of the ultimate consumer'. Within the supply chain of the oil and gas industry, upstream service is an essential part as it links directly to major petroleum company clients (oil operators).

Most firms engaged in the upstream services are highly specialised and own proprietary technologies. Large firms and industry leaders are also found to be highly concentrated within the upstream services segment. Because of their size and influence over buyers, suppliers and competitors, any changes within this sector would influence the level of service activity in general and alter the industry structure. Activities such as mergers and acquisitions at a time of low oil prices are considered a barometer on the health of the entire oil and gas industry.

Based upon some of the concerns addressed above, the priority of this empirical study was to focus on the upstream offshore oil and gas service sector.

1.2   Research rationale

1.2.1   Typology of the business environment

Strategic management theories and practices stress that understanding the business environment is essential in every organisation's life. Early in the 1950s, organisational theorists started to investigate organisation–environment interaction and found that the views of managers play a central role in learning about the environment (Tung, 1979). Managers are encouraged to become more responsive to the dictates of the external environment and are required to scan and assess environmental conditions when making strategic decisions. Environmental assessment implies identifying and evaluating how and why current and projected environmental changes affect or will affect the strategic management of an organisation (Fahey and Narayanan, 1986). Assessment attempts to investigate what key issues are presented by the environment and what the implications of the issues are for the organisation. Accurate assessment of the environment by managers may help bring about more effective strategies and thereby higher performance for long-term success (Downey et al., 1975; Hambrick, 1982; Daft et al., 1988; Hegarty and Tihanyi, 1999).

The term 'business environment' in the context of this study refers to the external environment. The external environmental influences include the remote environmental factors, which are normally addressed as political, economic, social and technological (PEST) factors; forces which drive change in industry structure and competition (Porter, 1985); and the critical operating factors for which firms compete. Porter (1980b) further develops a theoretical framework of using competitive strategy as a technique for analysing industries and competitors.

According to Johnson and Scholes (1999),

Managers face difficulties in trying to understand the environment. 'The environment' encapsulates many different influences;

the difficulty is making sense of this diversity in a way that can contribute to strategic decision making. Listing all conceivable environmental influences may be possible, but it may not be much use because no overall picture emerges of really important influences on the organisation.

Some of the macro-environmental influences are commonly seen as important to organisations. Examples of such environmental influences may include government action and restructuring, economic conditions, social culture, technology, ecology, demographics, labour market, capital markets and suppliers.

Strategies are affected by such influences (Johnson and Scholes, 1999). For instance, cultural factors may affect managerial style and the applications of strategy in an Asian environment. Especially in China, social networks have strong impacts on managers' decisions and hence affect strategic effectiveness (Warner, 2000). *Guanxi* (a personal relationship) has proved to be very important in securing business success as it can increase trust (Tan, 2001) in this country.

Family-based decision structures with extended business networks, conglomerate-style diversification, forward and backward vertical integration and acquisition emerge as predominant strategies of ethnic Chinese firms. Managers need to adapt to regional or local culture within their own or another country.

Nonetheless, environmental forces that are especially important for one organisation may not be the same for another; and, over time, their importance may change (Johnson and Scholes, 1999). A multinational corporation might be especially concerned with government relations and understanding the policies of local governments, since it may be operating plants or subsidiaries within many different countries whose political systems vary from each other. An oil and gas-related company, in particular, is more likely to be concerned with its technological environment that leads to product or service differentiations. In this sense, environmental analysis involves efforts to identify key issues.

Very recently, service companies (Sembwang, 2009; SembCorp, 2009) defined key external environmental factors including commodity price volatility, foreign exchange movements, the credit crunch, global economic recession, geopolitics and the regulatory environment.

Despite most literature on organisational or strategic management theories introducing the concept of the environment, comprehensive analyses or empirical studies of environmental characteristics are limited. Tung (1979) argues that a major obstruction has been how best to describe and conceptualise organisational environments. Most commonly, general environmental uncertainty reflects the nature of the business environment. Integrating the perspectives on organisational uncertainties has been continuously developed.

Some strategy researchers emphasise uncertainties are related to market demand for products or services, product and process technologies, the availability of critical inputs, and strategic actions by competitors and potential entrants (Miller, 1992). Other previous research (Daft et al., 1988; Tan and Litschert, 1994) uses environmental dynamism, complexity and hostility to measure the environmental uncertainty. Mintzberg et al. (1998) suggest that the organisational environment could be static or dynamic, complex or simple, hostile or favourable. Thompson (1967) emphasises that the priority for an organisation is to deal with the uncertain eventualities of the environment, particularly those of the task environment (Dill, 1958). The reliability of an instrument for measuring manager perception on the business environment is still to be developed and tested. Research on managerial perception of the business environment remains an important theoretical and empirical task.

1.2.2   Development of strategy concepts

With regard to strategy, the literature can be divided into two areas: one deals with the concepts of strategy in the East and the other focuses on strategy development in the West. Of these two areas, work conducted by Western strategists on strategy development is far more extensive. Much of this work is about the content of strategy, both in terms of conceptual frameworks and empirical evidence. Exact definitions of the strategy concept differ between Western and Eastern strategic theorists or strategists. However, what is never in question is the key attribute of strategy: the means of achieving a strategic goal. Basic doctrines about strategy – for instance, its importance to a firm's long-term success, the components of strategy and levels of strategy – receive almost universal acceptance (Hawkins,

1995). In a Western context, a large amount of work has been carried out in some key areas such as the strategy types an organisation can pursue and the organisational levels at which these strategies should be implemented.

Strategy is formulated not only by external environmental forces but also internal environmental factors. The distinctive strategic analysis is in relation to an understanding that needs to be considered in terms of the resource base and competences of an organisation (Johnson and Scholes, 1999). This refers to the so-called internal environment analysis (McNamee, 1992). In analysing the internal environmental factors, a company can be evaluated by investigating issues pertaining to organisational behaviour, organisational structure and management effectiveness. The behavioural analysis encompasses an understanding of missions, goals or objectives, organisational culture, leadership and strategy. Organisational structure includes informal and formal arrangements. The strategy may be appropriate but this may be of little use if the structure is wrong and/or the management is ineffective.

Researchers describe different ways to distinguish the nature of strategy. Generally, strategies can be compared in ways such as formal and informal (Alexander, 1990), tangible and intangible (Mintzberg, et al., 1998), realised and intended (Johnson and Scholes, 1999), and implicit and explicit (Pearce and Robinson, 1997). It was not until the late 1980s that modern Western strategic management and strategic theories were introduced into China at an academic level and were little applied in business practice. In fact, *The Art of War* by Sun Zi has been regarded as the most significant classical strategic thinking in East Asia (Chen, 1995). Therefore, within the context of this study, the strategy framework has been generated based upon both Western and Eastern strategic philosophies.

In most large organisations, their corporate strategies and business strategies are usually different; but in some organisations, especially small businesses, their corporate-level and business-level strategies may be the same. In each incidence, it is important to be clear about the basis of strategic options at a business level (Johnson and Scholes, 1999). Miles and Snow's (1978) typology and Porter's (1980b) taxonomy are two of the most widely accepted conceptual frameworks on business strategy (Carter et al., 1994; Slater and Olson, 2000).

Miles and Snow (1978) propose that organisations develop relatively enduring prototypes of strategic orientations in line with characteristics like the range of products and markets, technology solutions, desired growth pattern and attitude toward change. Researchers such as Hambrick (1982) criticise the Miles and Snow typology on the grounds that it is not the most elaborate framework that could be chosen. Researchers (Wright et al., 1990; Croteau and Bergeron, 2001; Parnell et al., 2000) have not only attempted to classify business strategies into typologies but also studied more effectively the relations between strategy and other variables such as performance. A common observation is that the more specific the type of business strategy adopted by an organisation, the better the organisational performance.

Porter (1980b, 1985) emphasises that the basis of generic business strategy is how customer or client needs can be best met, usually through achieving a certain competitive advantage. A competitive position is the basis on which a business might achieve competitive advantages in its market (Johnson and Scholes, 1999). A number of empirical studies have been conducted to test the validity of Porter's (1980b, 1985) generic strategies. Great attention has also been devoted to analysing generic strategies and competitive positions associated with organisational performance (Yamin et al., 1999).

Although business strategy theory has largely been developed based on research into organisations that operate in a Western business environmental context, little or nothing is known about the strategic solutions needed for oil and gas service companies to survive and prosper long term in the East Asian business environment. Particularly, business strategies adopted by such companies operating in Asia have not been written about in any great depth.

Strategists need to acquire information about the business environment in order to make strategic business decisions. In this field of study, there is still a need to develop concepts and ideas in order that a framework linking strategy with its business environment can be used as a tool for analysis of an industry such as oil and gas. Many previous general studies have been conducted concerning world oil markets. However, most research has been approached from the operating companies' perspective while much less attention has been paid to the service and support sector. Previous researchers have carried out studies in the field of assessing the business environment and business strategies, as well as evaluating the relationship between the two variables. This theoretical framework was employed as the bedrock of this study.

In short, based on a broad range of literature review, a research gap left by previous research work was identified. In order to narrow the existing research gap, three objectives emerged to guide this study.

1.3 Research objectives and propositions

As discussed above, a general aim leading this study examines the common trends concerning strategic orientations across the oil and gas service organisations in an Asian business environment. The theoretical substratum of this research is to evaluate the business environment and investigate business strategies adopted by indigenous and foreign service companies operating businesses in the selected East Asian countries. In order to achieve the research aim, three research objectives were established as follows:

1 To provide an understanding of the dominant business environmental factors which affect the oil and gas service industry in China and, to a lesser extent, in Singapore and Malaysia.
2 To investigate strategies adopted by oil and gas service companies in response to the business environment(s) in which they operate.
3 To evaluate, in the light of empirical evidence, the reliability of the assumptions drawn from the literature of strategic theoretical frameworks based upon Western business practice, and applied in a non-Western business environment like Asia.

The business environment can be evaluated from three aspects: objective, perceived and enacted. In this study, the objective business environment (Dill, 1958) refers to the measurable reality such as oil-related economic indicators, factors relevant to industrial and technological aspects, and government regulations and policies. The objective business environment will be dealt with as the industry background context in Chapter 2.

The perceived business environment (Mintzberg et al., 1998) implies managerial perceptions on the nature of the business environment. The investigation on the perceived business environment formulated the first objective of the study.

One of the research intentions was to categorise the companies in terms of five basic business strategic orientations, namely, balancer, analyser, defender, prospector and reactor (Miles and Snow, 1978; Parnell et al., 2000). The second objective was to investigate these strategies adopted by service companies. Business strategy options, in this research context, are regarded as the enacted business environment (Weick, 1979).

Next, the correlation between the perceived business environment and strategies adopted is examined. In the light of evidence gathered, the third research objective was to provide the analytical outcomes for evaluating strategic frameworks generated in Chapter 3.

Furthermore, ten propositions were proposed for the execution of the empirical study:

- Proposition 1: The six environmental sectors – technology, regulation, economics, customers, suppliers and competitors – can be defined as the key task environmental factors for the growth of service businesses in East Asia.
- Proposition 2: For oil and gas service companies that operate in East Asian countries like China, Singapore and Malaysia, the nature of the business environment will be perceived to be uncertain.
- Proposition 3: Oil and gas service companies' executives in East Asia perceive that the business environment in which they operate will be dynamic, complex and hostile.
- Proposition 4: The perceived environmental uncertainty will be associated with the perceived environmental complexity, dynamism and hostility.
- Proposition 5: The perceived environmental uncertainty will be associated with the influences of the task environmental factors.
- Proposition 6: The level of perceived environmental complexity, dynamism and hostility will be associated with the influences of the task environmental factors in the oil and gas service sector.

- Proposition 7: For oil and gas service organisations operating in East Asia, managerial perceptions on their business strategies will vary.
- Proposition 8: For oil and gas service organisations operating in East Asia, managerial perceptions on the business environmental uncertainty will be associated with their strategic orientations.
- Proposition 9: There will be relationships between the perceived business environment and strategic performance for oil and gas service organisations operating in East Asia.
- Proposition 10: For oil and gas service organisations operating in East Asia, strategic performance will be associated with their business strategic orientations.

The investigation of Propositions 1–6 is concerned with the first research objective. By evaluating Proposition 7, the second research objective can be achieved. The study seeks to fulfil the third research objective through the examination of Propositions 8–10.

1.4 Research methodology

Because the nature of the study required exploration of business environment and strategy, considerable use was made of qualitative research methodology. The empirical research was progressed by employing a multi-method approach. Joint methods of data collection, coding and analysis of data were the fundamental operations of this study. It involves a formal questionnaire survey addressed to senior management. A post-questionnaire survey, interview survey and telephone survey had been applied during the primary data collection process.

1.4.1 Questionnaire construction and research variables

The Survey Questionnaire (Appendix) is divided into three distinct sections, each dealing with one of the following issues: Section One, Background, together with Section Three, which comprises the Strategic Orientation and Strategic Performance parts, were used to survey the facts of organisations. Section Two, The Business Environment, which consists of the Environmental Sectors part and the Nature of the Business Environment part, was used to assess the attitude of managers or senior executives.

There were a total of 141 variables developed for this study, covering four areas: firm background, managerial attitude on the external business environment, managerial opinions on organisational strategic direction and managerial perceptions of organisational performance.

Dependent variables and independent variables are also encountered in this study. Managerial attitudes, opinions and perceptions were defined as the dependent variables because they are the criteria that had to be predicted or explained. The independent variables encompass organisational information such as company size, year and ownership types because these are the expected factors that influence the dependent variables. The categorisation into dependent and independent variables can been seen in the survey questionnaire.

1.4.2   Measurement and scales

The primary research developed the key concepts (dependent variables) to be investigated: the perceived business environment, business strategy and strategic performance. These three concepts were assessed by using a 111-category scale. Their mode of measurement is presented in the survey questionnaire.

Questions pertaining to the environmental sectors were based on a five-point Likert scale (which was advanced by Likert in 1932). For the assessment of the importance and impact of the environmental factor variables, respondents were required to indicate the extent to which six task environmental sectors, namely, economic, regulatory, technological, customers, suppliers and competitors, had affected their business over a defined period of the last five years. To indicate how important these sectors were for the growth of their business, individuals chose from five alternatives: not important, slightly important, important, very important and the most important. To indicate how the impact of these sectors affects the growth of their business, the five options were: non-existent or very weak, weak, moderate, strong and very strong. From very positive to very negative, the researcher also assigned scores of 5, 4, 3, 2 and 1 to the alternative answers relating to one attitudinal object. The weight of 5 was assigned to the response which was strong agreement indicating the most favourable attitudes on the statement.

Three environmental dimensions, viz. complexity, dynamism and hostility were defined as the measures for assessing the environmental sectors at a task environment level. The senior executives' views of and attitudes toward the environmental factors were measured by using a 45-item scale developed by the author. The questions were arranged as multi-item scales corresponding to the six environmental sectors. Each of the environmental items indicates the degree of environmental complexity, hostility and dynamism, and in turn, the level of the environmental uncertainty. The respondents were asked to rate the degree to which they agreed on various characteristics of the environmental sectors.

The answers on the perceived complexity and hostility were measured by a series of 7-point bipolar rating scales (Zikmund, 2000). Bipolar adjectives anchor the right and left of the scale, with 1 indicating that the respondent most strongly agrees with the left assessment; 7 indicating that the respondent most strongly agrees with the right assessment; and 4 showing that the respondent felt that, for his organisation, the situation was midway between both. Similarly, the environmental dynamism was examined by perceived predictability of these six environmental sectors. The answers were measured by using a 7-point numerical scale, with 1 indicating very predictable, 7 indicating very unpredictable and 4 indicating a neutral situation.

Questions of the Strategic Orientation section comprised multi-item scales including Miles and Snow's typology, Porter's taxonomy and Bowman's strategy clock. Hence, the data could be transformed in a category form showing various strategic classifications.

A 7-point Likert scale was utilised to assign strategic orientations and competitive strategies, with 1 indicating strong agreement and 7 indicating strong disagreement. Questions were randomly arranged to eliminate question-ordering bias. In total, 28 questions pertaining to business strategies were developed. Each of the items was designed for assigning different types of business and competitive strategies.

Firstly, the construct of Miles and Snow's business strategy was measured by examining five types of strategic orientations, namely, defender, prospector, analyser, balancer and reactor. The appropriate measures were constructed to assess the range of product or service domains, product-market strategies, the attitude toward change and the approach to managing change. The methods advanced by previous strategic researchers (Parnall et al., 2000) were employed to classify business strategies. A total of 20 questions were designed to deal with the variations of managerial perceptions in business strategies.

Among these, one question was related to whether they had clearly articulated business strategies in supporting the assignment of strategic categories.

Secondly, eight questions were used to assess competitive strategies and two of these questions pertained to whether organisations were aware of competitive strategies in supporting the associated assignment. Based on Porter (1980a, 1985), four generic strategies, namely, low-cost, differentiation, hybrid and no-purpose, were selected and examined in this study.

Furthermore, competitive advantage is the underlying concept of generic strategies and all organisations are in a competitive position in relation to each other as they compete either for customers or for resources (Porter, 1985; Johnson and Scholes, 1999). Because of the characteristics of the service sector within the oil and gas industry, firms compete mainly for customers (i.e. operators).

In order to understand an organisation's strategic position within which it attains competitive advantage, Bowman's strategy clock (Johnson and Scholes, 1999), which is advanced by Faulkner and Bowman (1995) as a Customer Matrix, had been developed further in this study.

Two dimensions of the strategy clock were defined as firstly, the perceived customer added value (PCAV, which is also conceptualised by Faulkner and Bowman as perceived use value) and secondly, the offering price. The first dimension, PCAV, refers to the senior managerial perceptions on the degree of the value that their businesses create to satisfy customers' needs. For oil and gas service organisations in East Asia, it was assessed by managerial perceptions on five variables: quality valued by customers, technological reliability, safety performance, speed of response to customers and price that customers are willing to pay. The second dimension refers to the level of price offered by service firms for their products or services. It was measured by managerial perceptions on the level of price charged.

A 6-item scale (i.e. six questions included) was used in an attempt to depict the strategic competitive positions for the participating organisations. For the first five items, a 5-point semantic differential scale was used with 1 indicating very low, 3 indicating a moderate and 5 indicating very high position. For the price that customers are willing to pay, 1 indicates much lower than that a service provider charges, 3 indicates the same as the service provider offers and 5 indicates much higher than the service provider offers. The weights of 1, 2, 3, 4 and 5 shown in parentheses were not printed on the questionnaire.

To evaluate strategic performance, there were 20 questions developed covering overall organisational activity areas such as finance, operations, marketing, human resource management and other strategic aspects. This study employs a subjective measurement (Dess and Robinson, 1984) that calls upon managerial perceptions. The reliability of this self-reporting approach has been proved by various studies (Tan and Litschert, 1994; Luo and Tan, 1998; Luo and Park, 2001). A 7-point interval scale was drawn from work by Ramanujam and Venkatraman (1987), with 1 indicating much less or worse and 7 indicating much more or better.

1.4.3   Sampling

Alexander (1990) says research indicated the best rationale is not to post questionnaires to firms that have been operating for less than seven years in a European context. A similar study (Wilson, 1992) indicates that eight years are an average for a new venture to reach breakeven with some ventures taking up to 12 years. In the context of the oil and gas service industry, such findings appear not to be valid. Secondary data show that some new firms may be set up as replacements for older ones. In this case, the replaced leadership might continue business by combining their experience with strategies already in use. Even a five-year-old, reborn and relabelled service firm could generate strategies successfully to enable the firm to outperform competitors. Theoretically, a strategy should be made in a long-term sense, normally a minimum of five or ten years. In this study, a firm would be eligible for selection if it had been operating for a minimum of five years in the selected countries. As a result, firms operating for fewer than five years were excluded from the results analysis on business strategies and associated strategic performance.

1.4.4   Data collection

In order to collect primary data, a survey questionnaire approach was utilised. The collected data were based on a questionnaire survey sent to the senior managers of oil and gas service companies operating in China, Singapore and Malaysia.

Five hundred questionnaires were distributed to the senior management respondents via post (mainly), email and fax between late 2001 and early 2002; then follow-up phone calls and emails constituted an attempt to achieve the proposed response target. Each of the packets mailed contained a covering letter with the correspondent's photograph on it, the questionnaire and a self-addressed envelope for return.

By August 2002, a total of 108 completed questionnaires had been returned by managers involved in operating businesses in China, Singapore and Malaysia, with a response rate of 21.6 per cent. The response rates from China, Singapore and Malaysia were 18, 23 and 14 per cent respectively. Of those, 98 completed questionnaires were usable for the final primary data analysis. For reasons of confidentiality, the names of respondents and companies were classified in a code term. The associated details of these companies can therefore not be identified by anyone except the author.

1.4.5   Data analysis

The focus of the analysis was to justify the relevance and validity of the data obtained and to discuss the extent to which the theories that had been started with have been developed as a result of the research. Three interrelated procedures form the analysis stage: descriptive analysis, univariate statistics and bivariate analysis (Kerr et al., 2002). The aim of using descriptive statistics was to provide interpretation of raw data in a format in which responses or observations can be described and understood easily. Univariate analysis allowed assessment of the statistical significance of various hypotheses about a single variable. Bivariate statistics was utilised to measure the difference or the association between two variables at a time (Zikmund, 2000).

In this study, most data were measured at a nominal and ordinal level. Ordinal data gave the first consideration of adopting the median as the measure of central tendency. The histogram graph was also drawn up to examine skewness. The general picture obtained was skewed and non-normal distributions. This finding had a major influence on the choice of non-parametric statistical techniques to be used for examining the proposed propositions, assumptions and hypotheses.

Selection of the various statistical techniques was based on their relevance to the research objectives. Major techniques adopted for data analyses and the associated research aspects are summarised as follows:

- To analyse the background information on company profiles, industry segment and business activities in East Asia, frequency distributions were used.
- Histogram graphs were drawn up to examine the skewness of some variables; the Kolgomorov–Smirnov goodness-of-fit test was performed to check for normality.
- To evaluate the significance of each of the environmental sectors, the one-sample chi-square test was computed.
- To evaluate differences in medians among the managerial perceptions of the importance or impact of the six task environmental factors, Friedman's tests were conducted.
- To test cognitive coherence between the perceived business environment uncertainty and the perceived dynamism, hostility and complexity, the Spearman correlation and chi-square tests were used.
- To evaluate the relationship between the two variables among the three dimensions of the perceived business environment, business strategic orientations and strategic performance, the Spearman correlation and Crosstabs with two-way contingency table analysis and chi-square tests were employed.
- To test the differences of the perceived environmental uncertainty and strategic performance by strategic groups, the Kruskal–Wallis tests were applied.
- A number of box plot graphs and scatter diagrams were also applied to highlight the pattern of differences and correlations.

1.5   Significance of the research

This book investigates strategies chosen by the oil and gas service companies operating in China, Singapore and Malaysia, and the extent to which indigenous business practice can be aligned with that of Western industrialised nations. It provides an analytical view of the reliability of strategic theoretical frameworks based on Western business practice, but applied in a non-Western business environment like Asia. The book also highlights the most recent trends against the empirical research findings, including the impacts of the global credit crunch on the oil and gas service industry in the turbulent times of 2008–9. As such, this research has made contributions to the previous work in four respects.

First, this research deals with the existing literature on Western strategic models, frameworks and theories, and examines them in Eastern contexts. The expected outcome is that the work will lead to an understanding of whether the Western strategic theories can be adapted for successful application in Asia.

Second, it has been argued that it is very challenging for any researchers, especially those who are familiar with Western theories and contexts, to conduct empirical work in Asia. This study developed a methodology for data collection and analyses in Asian countries like China, Singapore and Malaysia. The methodology generated for this study can be utilised by academics, business researchers or analysts, or other experts to conduct research in a similar industry or country context.

Third, this study gathered empirical evidence from the questionnaire survey of selected firms operating in Asia. It is expected that the findings on the impact of the business environment and significance of adopted strategies in Eastern contexts will add empirical evidence to extend the existing strategic management and Asian management literature.

Finally, as the oil and gas service firms under investigation are all companies operating in Asia, the findings from this research will enrich our knowledge of both the petroleum and wider energy service industries in general and the service sector in particular. The research findings will make a useful contribution for corporate senior management in generating effective strategies to increase their ability to survive long term in the Asian market. Taken within this context, the book is also of interest to academics, managers and other experts specialising in Asian studies and/or the oil and gas industry.

1.6   Organisation of the book

Following this chapter, Chapter 2 examines the oil and gas service industry in China, Singapore and Malaysia, with a focus on the objective facts of dominant environmental factors. Six environmental factors are identified: the oil and gas-related economics, technology, political and regulatory factors, oil service companies and their competitive activities, suppliers' resources for the oil service companies, and regional oil operators (customers).

Chapter 3 develops a selected review of the literature on frameworks of the business environment. It also deals with the major thinking and research on strategy, which is written comparatively and chronologically. The variables of business environmental dimensions and business strategic characteristics as well as the overall indicators of strategic performance are discussed. Some of the relationships among these variables are explored. It also develops theoretical frameworks and ten major research propositions drawn from the literature review in order to test the data gathered from the empirical investigation.

Chapter 4 demonstrates empirical findings on the oil and gas service sector in China, Singapore and Malaysia. It contains the background context of the oil and gas service companies involved in this study and managerial perceptions on the business environment in which they operate. The differences or similarities of managerial perceptions on the business environment are presented.

Chapter 5 examines and categorises different business strategies deployed by the participating service organisations. It reveals competitive strategies and strategic positions for the service companies operating in the East Asian market. The chapter also compares differences and similarities of strategic directions and organisational performance across the three selected East Asian countries, China, Singapore and Malaysia. In order to examine strategic effectiveness, it also contains a comparative analysis on the strategic outcomes for service companies that adopt different business strategies.

In Chapter 6, the completed analysis measures the correlation between the types of strategies pursued by the participating organisations and strategic performance, by taking account of the perceived environmental uncertainty. This is carried out in terms of three aspects. First, it discovers the alignment between the perceived uncertainty and strategic options. Next, it establishes the relationships between strategies and the associated strategic performance. Then, the correlations between the perceived environment and strategic performance are explored.

Chapter 7 discusses the most recent trends of the oil and gas service industry in the light of the findings from the empirical research.

It reviews the significance of dominant business environmental conditions and strategic trends within the energy service sector under scrutiny. In particular, the chapter highlights the impacts of the global financial crunch on the service sector in China, Singapore and Malaysia during the turbulent times of 2008–9.

Chapter 8, the concluding chapter, summarises the results of the empirical research and discussions on the recent strategic trends of the oil and gas service industry in East Asia. It is hoped this book proves useful both to business strategists (or those responsible for formulating business strategies) and to researchers in academic, industrial and government spheres. Energy service organisations with particular interests in East Asia should find that the conclusions can assist them in devising profitable strategies for their businesses in the region. The significance of the research findings in this book is, the author believes, important for both Western and Eastern senior executives who wish to enhance their businesses effectively so that they may survive and prosper long term in Asia.

Note

1. In this book, East Asia includes China, Hong Kong, Taiwan, Vietnam, Indonesia, Malaysia, Thailand, the Philippines, Singapore, South Korea and Japan.

2
The Energy Service Industry in East Asia

This chapter seeks to provide an understanding of the service sector within the oil and gas industry in East Asia, with an emphasis on China, and to a lesser extent Singapore and Malaysia. It examines the dominant business environmental factors. These are defined as the petroleum-related economics, technology, regulations or politics, the upstream service companies and their competition, suppliers to the upstream service companies and oil operators (customers). Each environmental factor is dealt with in two dimensions: identifying and presenting the key issues of that factor; and comments on the impacts of these environmental factors. The volatile business environment during 2008–9 and the impact of the world financial crunch that hit its peak in late 2008 will be discussed in Chapter 7.

2.1   Generic context and background

Southeast Asia's economic problems of 1997 are widely documented; however, the situation had become brighter by 2000. Malaysia was fastest in terms of its economic recovery, while Singapore, as a powerful economy, continued to grow. In the past decades, China has been the engine of growth for Southeast Asia and its stable market climate has been attractive to foreign investors.

In spite of the Asian financial crisis in the late 1990s, China has been able to sustain a sound economic performance featuring rapid growth, low inflation (about 1 per cent in 2001) and highly profitable business opportunities. Despite the creeping worldwide recession, GDP (gross domestic product) grew by 8.2 per cent in 2000; the trend continued with an increase of 8.3 per cent in 2001 while 2002 recorded a 9.1 per cent GDP growth rate (*The Economist*, 2003).

Figure 2.1 shows China's annual GDP growth rates for 2003–5 were recorded at around 10 per cent in real terms. Despite efforts to cool the overheating economy, the officially recorded GDP growth rate was 11.4 per cent in 2007. However, its growth rate was hit by the global economic meltdown in 2008, when GDP growth dropped to 10.6 per cent. Even though the global credit crunch has hit China's economy hard, economic performance has remained relatively robust,with an expected GDP growth of 6.0–6.5 per cent for 2009, and the country is still outpacing most other nations (The World Bank, 2009; *The Economist*, 2009). Chinese economic growth is forecast to remain strong, climbing to 7.0 per cent by 2010 (*The Economist*, 2009).

Malaysia returned to strong economic growth following the deep recession caused by the Asian financial crisis of 1997 and 1998. The country had an annual average growth rate of 5.4 per cent between 1994 and 1998. Due to the impact of the 1997 financial crisis, GDP suffered a 7.4 per cent decline in 1998.

In response to the dynamic environment conditions, the Malaysian government adopted a reflationary economic policy to boost exports and the domestic market. Spending on capital investment was

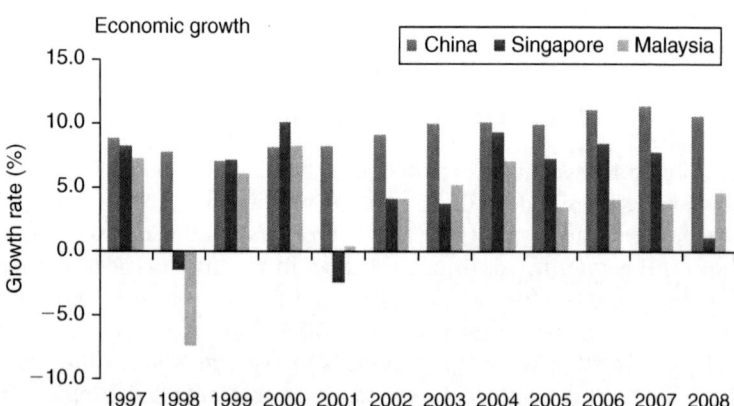

Figure 2.1   GDP in China, Singapore and Malaysia, 1997–2008
Source: National Bureau of Statistics of China (2009); Statistics Singapore (2009); Department of Statistics Malaysia (2009).

restricted. As a result, Malaysia made a quick economic recovery in 1999 from what had been its worst recession since independence in 1957 (EIA, 2001). It experienced GDP growth of 6.1 per cent in 1999 and a strong 8.3 per cent in 2000. This recovery was credited to rapid growth in exports, particularly of electronics and electrical products, to the United States, Malaysia's principal trade and investment partner. The country's economy slowed dramatically in 2001 due to the weakening demand for Malaysian exports, which were, at the time, associated with the economic slowdown in the United States (ibid.). In 2002, its GDP grew only 4.1 per cent (*The Economist*, 2004). Annual growth reached 5.2 per cent in 2003. Malaysia posted 7.1 per cent GDP growth for 2004 (Department of Statistics Malaysia, 2009).

The growth trend continued into 2008, but in 2009, its economy was hit by the global credit crisis and a sharp adjustment of oil prices. A major challenge for Malaysia will be to try to recover from the adverse effects of this global economic meltdown. *The Economist* (2009) expects real GDP growth to average 1.7 per cent a year in 2009–13, a low rate by historical standards.

Strategically located at the entrance to the Strait of Malacca, Singapore has become one of the most important shipping centres in Asia. Its prosperity and economic health are attributed to its port facilities – the world's busiest by shipping tonnage. The island state is blessed with a highly developed and successful free-market economy and a remarkably open business environment. Exports, particularly in electronics and chemicals, and services are the main economic drivers. Well-developed infrastructure, a transparent legal system and skilled workforce have helped Singapore to establish itself as one of the top three global oil trading and refining hubs (US Commercial Service, 2009).

GDP in Singapore grew 10.1 per cent in 2000, but this fell to a negative 2.4 per cent in 2001 largely because of weaker global demand. Its economy recovered in 2003–4 and performed strongly in the following years, achieving 9.3, 7.3, 8.4 and 7.8 per cent in 2005, 2006, 2007 and 2008 respectively (Statistics Singapore, 2009).

The Economist (2009) says that Singapore's economy will contract sharply in 2009, reflecting its exposure to the global economic slowdown. Growth is predicted to increase steadily from 2011 as demand for the country's main export markets improves. However, growth rates may not return to the high levels that have been recorded in the past few years.

Petroleum product demand and consumption throughout the Asia-Pacific region have increased dramatically with rapid economic growth. Among the top five energy-hungry countries of the world, four are located in the Asia-Pacific region, with China leading in energy consumption (Ng, 2009).

China has lately accounted for half of global energy consumption growth. In Malaysia, to meet increasing demand, the offshore gas supply has grown rapidly; almost 20 per cent of total supply is imported. Energy consumption has also increased due to the development of business and industry in Singapore. The country is experimenting with various forms of alternative energy – solar, fuel cells, biodiesel and even hydrogen/natural gas hybrid systems (US Commercial Service, 2009).

According to the National Bureau of Statistics of China (2009), China's energy consumption experienced a growth rate of 7.8 per cent in 2007 and 7.8 per cent in 2006. On the back of falling factory output and exports triggered by the global financial crisis, energy consumption in China registered a 4.0 per cent growth rate to the equivalent of 2.85 billion tonnes of standard coal in 2008. That was the smallest increase since 2003 when China started releasing energy consumption data. China relies largely on coal as the primary energy resource (Figure 2.2) despite the volatile trend in supply.

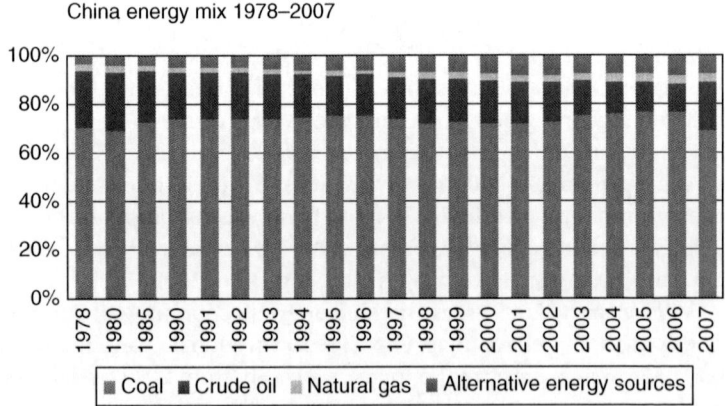

Figure 2.2 Energy consumption in China, 1978–2007
Source: National Bureau of Statistics of China (2009).

China has been trying to cut energy consumption for some time as it aims to improve its energy efficiency and targeted a 20 per cent cut in consumption in the five years to 2010. Energy use for each unit of gross domestic product (GDP) was trimmed by 4.59 per cent in 2008 (US Commercial Service, 2009).

According to the National Bureau of Statistics (2009), gas consumption for 2008 rose by 10 per cent to 80.7 billion cubic metres (bcm). Beijing has encouraged the use of this fuel as a cleaner alternative to thermal power and other fossil fuels. Crude oil consumption in the same year also increased 5.1 per cent to 360 million tonnes, though this was the smallest growth for three years.

2.2 Petroleum industry review

2.2.1 China

Before 1939, over the period that Japan invaded China, very little oil exploration was possible. The oilfields in the northern part of Shanxi province were targeted but with little result. In 1939, the Yumen oil field was discovered. Between 1939 and 1949, Yumen produced 524,000 tonnes of oil in total. In 1949, when the People's Republic of China was established, total crude oil production in China only reached 120,000 tonnes a year. At the time, Mobil became the first foreign company to conduct oil reserves surveys and explorations in the country (Scottish Enterprise, 2002). Following the discoveries of the Daqing and Shengli oilfields in 1959 and 1962 respectively, a number of new oilfields were found. Since then, China's oil production has increased rapidly. The details of the major post-1949 oil discoveries in China are shown in Table 2.1.

A significant breakthrough by the Chinese petroleum industry took place in the 1980s, at a time when production in China's major onshore oilfields began to decline. The Chinese government stepped up its exploration in both western onshore regions (Tarim, Juggar and Qaidam basins) and offshore.

As a result of restructuring, the three largest oil companies were established during the 1980s. The China National Offshore Oil Corporation (CNOOC) was formed in 1982, while the China National Petroleum Corporation (CNPC) and the China Petrochemical Corporation (Sinopec) were established in 1988 and 1983 respectively (CNPC, 2009). By the late 1980s, most of China's sedimentary basins had been

Table 2.1   China's post-1949 major petroleum activities

Date	Oilfield discoveries and operations
31 October 1955	Karamay oilfield with Exploration-1 well striking commercial oil in the Junggar basin of Xinjiang
13 September 1958	Qinghai's oilfield with Lenghu district Dizhong-4 well striking commercial oil
26 September 1959	Daqing oilfield, a world-class giant oilfield with Songji-3 well striking commercial oil in the Songliao basin in Heilongjiang
16 April 1961	Shengli oilfield with Yinghua-8 well achieving high-production oil flows in Dongying, Shandong Province
20 December 1964	Dagang oilfield with Gang-5 well producing significant oil flows
1 August 1969	Construction of Jianghan oilfield commenced
9 September 1969	Liaohe oilfield located with Xing-1 well in the Liaohe basin
20 April 1970	Jiling oilfield development began
27 June 1971	Changqing oilfield discovered with Maling Ling-9 well striking commercial oil flow in Maling, Gansu Province
8 August 1971	Henan oilfield was discovered
4 July 1975	Huabei oilfield found with Ren-4 well striking commercial oil flow in Bohai Bay
7 September 1975	Zhongyuan oilfield found with Pushen-1 well in Dongpu, Henan Province
17 May 1977	Kekeya oilfield located in the Tarim basin of the southern Xinjiang area
2 March 1989	Turpan-Hami oilfield with Taican-1 well striking commercial oil in the Turpan-Hami area in Xinjiang

Source: CNPC (2009).

the subject of geological and geophysical surveys and many exploration wells had been drilled in the more prospective areas. China began to open onshore areas to foreign companies for exploration.

The Chinese oil industry has been controlled and developed by a variety of ministries in China. The indigenous oil service companies are basically derived from the state-owned oil companies (Table 2.2). The number of players in the service sector has been built up by

Table 2.2 Major industrial participants in China

Year	Participants
1949–55	Ministry of Fuel Industry responsible for the production and establishment of petroleum industry
1955–70	Ministry of Petroleum Industry responsible for oil exploration and development
1970–75	Ministry of Fuel and Chemistry Industry (FCIM) was derived from the ministries of petroleum, coal and chemistry
1975–78	Ministry of Petroleum and Chemistry Industry was formed to replace FCIM
1978–88	The era of the Ministry of Petroleum Industry
Feb. 1982	China National Offshore Oil Corporation (CNOOC) was established to be responsible for offshore exploration in association with foreign companies
1983	China National Petrochemical Corporation was formed to be responsible for downstream activities
1988–98	As a result of industry reform, the China Petroleum Corporation was formed to be responsible for onshore exploration and production, with partial government responsibilities; meanwhile, the Ministry of Petroleum Industry was dismantled
1998	The Chinese petroleum assets between China Petroleum Corporation and China National Petrochemical Corporation were reorganised. The restructured and today well known CNPC and Sinopec were established
Nov. 1999	PetroChina was formed
2000	PetroChina was listed on both the New York and Hong Kong stock markets
Apr. 2000	COOEC, a CNOOC offshore service arm, registered at Tianjin Industrial and Commercial Administrative Regulatory Bureau
May 2001	China Petroleum Engineering Group (CPE), a CNPC supply and services subsidiary, was formed
Sep. 2002	COSL, a CNOOC offshore service arm, was formed
Dec. 2002	CNPC Well Logging Company Limited was formed
Nov. 2004	CNPC Offshore Engineering Company Limited (CPOE), CNPC's offshore service arm, was formed

Source: CNPC (2009); Sinopec (2009); CNOOC (2009).

new entrants such as international service companies, increasingly independent state-owned or partially state-owned service companies, and private professional companies.

Before 1998, CNPC, Sinopec and CNOOC administered the Chinese petroleum industry on behalf of the government. They were awarded considerable coordinating and supervisory powers by the State Council. Nonetheless, as a result of the government's continuing market economic drive in 1998, the companies handed over all of their previous regulatory functions to the State Economic and Trade Commission (SETC), enabling central government to assume full regulatory control of the sectors. The companies adopted a market orientation with their attention focused mainly on commercial activities (Scottish Enterprise, 2002).

Since the 1990s, in order to streamline industry operations and management efficiency, the Beijing government has carried out a fundamental reform of China's petroleum industry. In particular, the three largest national oil companies, CNPC, Sinopec and CNOOC, were partially reorganised.

In a major restructuring process started in 2000, the two integrated companies, CNPC Group and Sinopec Group, were created from the assets of the former upstream company CNPC and the downstream company Sinopec. Their assets were redistributed, with upstream and downstream assets in the north and west of the country handed to CNPC, and those in the south allocated to Sinopec. CNPC then created a new company, PetroChina, to hold all of its operating assets, with CNPC acting as a holding company for minority interests and controlling the state's interest in PetroChina. The restructuring of CNOOC also resulted in the establishment of two major offshore service contractors, COOEC and COSL.

In 2000, shortly after the financial crisis in Southeast Asia and following the late 1997–99 oil price slump, PetroChina Limited, Sinopec Limited and CNOOC Limited were listed on the New York, London and Hong Kong stock exchanges. PetroChina was later also listed on the Shanghai stock exchange in 2007.

The offshore division of CNOOC was formed to facilitate the development of offshore exploration in conjunction with foreign oil companies. Many companies, including BP, Shell, Esso and others, offered considerable work programmes in return for the right to explore and produce oil and gas under new production sharing contracts (PSCs).

Initial activity focused on the Pearl River mouth basin, East China Sea and the Yellow Sea. Despite considerable efforts, results were generally disappointing. Some small oilfields were discovered and developed in the Pearl River mouth basin and the Beibu Gulf (north of Hainan Island), but the biggest prize was Arco's Yacheng-13 gas field off the southern coast of Hainan.

In the early 1990s, the blocks released in the Tarim basin were of little interest to foreign companies. But in the late 1990s, some apparently more attractive areas were offered to foreign participation for development and enhanced oil recovery projects. As a result, Kerr-McGee and Phillips in particular made significant oil discoveries in Bohai Bay and this stimulated further interest in exploration from foreign companies (Scottish Enterprise, 2002).

Until 2009, offshore activities remain dominated by CNOOC. However, this monopoly will soon be broken, as CNOOC's two major competitors, CNPC and Sinopec, will be granted the same right to conduct offshore exploration and production (E&P) activities and to establish cooperation with foreign interests (CNOOC, 2009).

2.2.2 Malaysia

In Malaysia, there has been substantial oil and gas activity for over 100 years. The Malaysian petroleum industry started when the Anglo-Saxon Petroleum Company, the precursor of today's Sarawak Shell, began to explore in the town of Miri, Sarawak, and made its first oil strike in 1910. This spurred further exploration.

By the 1950s, attention was turning offshore. It was in 1962 when oil was found in two areas offshore Sarawak. In Peninsular Malaysia, petroleum exploration activities commenced in 1968 with the first oil found in 1971.

Malaysia's Petroleum Development Act was enacted in 1974 and in the same year the National Oil Company of Malaysia, PETRONAS (Petroleum National Berhad), was established (Table 2.3). Its responsibility is to ensure that the nation's petroleum resources are developed in line with the country's needs. According to the Petroleum Development Act, PETRONAS 'has the entire ownership in, and the exclusive rights, power, liberties and privileges of exploring,

Table 2.3   Major petroleum activities in Malaysia

Year	Participants
1882	The earliest officially recorded oil find was in Sarawak
1910	The Anglo-Saxon Petroleum Company, the forerunner of the present Sarawak Shell, struck oil at Miri, marking the start of the Malaysian petroleum industry
1974	Parliament passed the Petroleum Development Act; incorporation and establishment of PETRONAS
1978	Incorporation of PETRONAS Carigali Sdn Bhd, the E&P subsidiary of PETRONAS
1982	PETRONAS Carigali made its first oil discovery, the Dulang oilfield, offshore Terengganu
1997	Malaysia International Shipping Corporation Berhad (MISC) became an offshore logistics arm of PETRONAS
2008	Commencement of oil production from the Kikeh field, Malaysia's first deepwater project

Source: PETRONAS (2009).

exploiting, winning and obtaining petroleum both onshore and offshore of Malaysia' (PETRONAS, 2009).

Since its establishment, PETRONAS has been manager and regulator of the country's upstream sector. More recently, it has also evolved towards being an international petroleum corporation.

In 1997, PETRONAS acquired a 29.3 per cent stake in the Malaysia International Shipping Corporation Berhad (MISC). In the same year, PETRONAS integrated PETRONAS Tankers Sdn Bhd into MISC, making MISC a 62 per cent subsidiary, an offshore logistics arm, of PETRONAS.

With a population of nearly 28 million people in 2008 (Department of Statistics Malaysia, 2009), Malaysia has since the 1960s developed her own infrastructure and adopted policies designed to support the development of domestic established companies. With relatively healthy oil prices and increasing demand (both domestically and internationally) for gas, upstream activity is on the rise.

Malaysia is currently focusing on development of various oil and gas fields offshore Peninsular as well as East Malaysia. As of May 2009, there are 88 fields in production in Malaysia, of which 61 are oilfields. Out of 183 gas fields discovered, 27 are in production with several more in development. In addition, there are 67 PSCs in operation, with 23 in Peninsular Malaysia, 21 in Sarawak and 23 in Sabah (PETRONAS, 2009).

Malaysia's deepwater reserves have doubled over the past five years. The most recent such discovery offshore Sabah is of significant importance to future offshore production activities in the country. The substantial capital expenditure required to monetise various discoveries has generated new service opportunities including sub-sea and gas transmission pipelines as well as oil and gas receiving terminals.

2.2.3 Singapore

The oil and gas industry has been an integral part of Singapore's economy since its oil-trading activities started in 1891. In the 1960s, Singapore strategically established itself as the region's leading oil refining and petrochemicals centre, as well as the logistics support basis for the Southeast Asia's oil and gas industry.

The industry contributed almost 5.0 per cent of Singapore's GDP in 2007. The government is currently focusing on playing a leading role in the global energy service market through innovative logistics solutions for refining, trading and logistics activities. Today, Singapore is one of the world's top three export refining centres (Singapore EDB, 2009).

Despite its neighbouring countries' rapid development in producing petrochemicals, Singapore remains a centre of major business activities, including acting as a regional hub for numerous service and supply companies. Many manage their regional operations stretching from the north-west shelf of Australia to the Middle East, and north to China and beyond.

Singapore seeks a future increasingly marked by globalisation. The country is positioning itself as the financial and high-tech hub in East Asia (McNulty, 2001; CIA, 2001) and is already the region's high-tech leader (DTI, 1996).

As already indicated, Singapore is significant in terms of oil and gas-related service activities, especially marine and offshore engineering (Singapore EDB, 2009).

Nearly 70 per cent of the international oil and gas suppliers manufacture in Singapore; local SMEs are mainly sub-suppliers providing manufacturing support and ancillary services to major suppliers (Ng, 2009).

The marine and offshore service industry in Singapore has enjoyed a rich harvest for some years, especially building drilling rigs and floating production units (FPUs). Today, it is the largest manufacturer of jack-up rigs, accounting for 70 per cent of the world market. It has 70 per cent of the global market for the conversion of floating production storage offloading (FPSO) units. Its ship repair sector also has a 20 per cent share of the world market (Singapore EDB, 2009).

2.2.4   Brief history of major service companies

The following is a brief introduction to the development of major state-owned oil companies' offshore oil and gas services (Figure 2.3).

In late 1999, whilst I was carrying out pilot fieldwork in China, the Chinese drilling companies, the China National Star Petroleum Corporation, the China Offshore Oil Northern Drilling Company and the China Offshore Oil Southern Drilling Company, owned rigs.

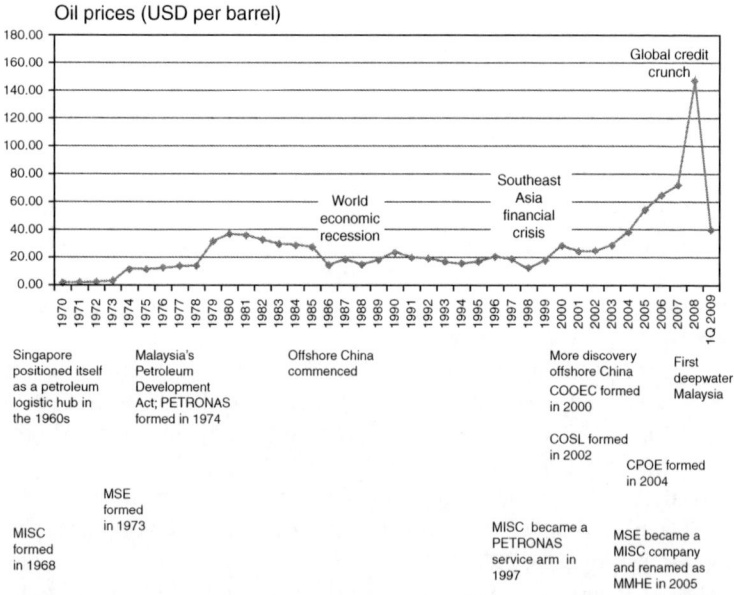

Figure 2.3   Volatile global oil economy versus emerging offshore service sector in East Asia

They operated in the East China Sea, the South China Sea and North China Sea (World Oil, 1999).

On 18 April 2000, with the approval of the former State Economic and Trade Commission (SETC), COOEC was established as a joint venture between five CNOOC subsidiaries, namely CNOOC Design Company, CNOOC Platform Manufacturing Company, CNOOC Maritime Engineering, CNOOC Bohai Company as well as CNOOC Nanhai West Company. In February 2002, COOEC was listed on the Shanghai stock exchange. By the end of 2008, CNOOC held 56.7 per cent of its shares.

By 2001, CNOOC was responsible for its own drilling, well and geophysical services through five subsidiaries:

- China Offshore Oil Southern Drilling Company;
- China Offshore Oil Northern Drilling Company;
- China Offshore Geophysical Company Limited;
- China Offshore Logging Company Limited;
- CNOOC Petrotech Services Company.

At the end of 2001, CNOOC incorporated these subsidiaries as a service company called CNOOCs to provide the same offshore services. Meanwhile, CNOOC also merged other two subsidiaries, the Offshore Oil Southern Shipping Company and the China Offshore Oil Northern Shipping Company, into one entity.

Having completed the restructuring process, COSL was officially established in September 2002. Two months later, COSL was listed on the Hong Kong stock market and in 2007 COSL was listed on the Shanghai stock exchange. By the end of 2008, CNOOC held 54.7 per cent of its shares.

Malaysia Marine and Heavy Engineering Sdn Bhd (MMHE) is a flagship for Malaysian heavy industry. Its establishment was associated with two organisations: Malaysia Shipyard and Engineering Sdn Bhd (MSE) and MISC.

In 1968, MISC was incorporated via a joint venture between the Malaysian government and a number of private entrepreneurs. MSE was incorporated on 18 May 1973.

PETRONAS acquired a 29.3 per cent stake in MSIC in 1997. A year later, PETRONAS increased its stake in MISC to 62.01 per cent by a merger between its subsidiary PETRONAS Tankers Sdn Bhd and MISC.

In 2004, MSE became a subsidiary of MISC. In May 2005, MSE was rebranded with a new name, Malaysia Marine and Heavy Engineering Sdn Bhd (MMHE). As a result, MMHE became a wholly owned subsidiary of MISC.

2.3   Key indicators

As outlined at the start of this chapter, the essential variables indicating the economic trends for the oil and gas service companies as a whole, or in part, include oil prices, E&P levels, reserve levels, expenditure of oil operators, well counts and rig counts. All of these point to the business conditions for energy service companies.

According to some drilling operators such as COSL (2009) and Noble Corporation (2009), the prices for oil and gas have historically been volatile and future price movements are unpredictable. These are mainly subject to a number of complicated global issues including:

- the supply of and demand for oil and gas;
- actions of or initiatives by the Organisation of Petroleum Exporting Countries (OPEC);
- other oil-producing countries' actions to control prices or change production levels.

The levels of exploration, development and production activities vary in different regions and these are reflected or measured by the capital expenditure budgets of oil and gas companies. The trends of service activity or rig counts are governed by the level of activity in exploration, development and production. Rig counts rise and fall with oil companies' budgeting and spending cycles (Noble Corporation, 2009; Ensco, 2009).

Trends in reserves stimulate future exploration and drilling activity. As a result, all associated service activities such as engineering, procurement and construction management (EPCM), topside fabrications, vessel support and helicopter services should increase accordingly (COHC, 2009).

In their annual reports, upstream service companies (Pride International, 2009; COSL, 2009; COOEC, 2009) point out that petroleum companies' spending on exploration, development and production is affected by factors such as:

- the current and expected prices of oil and natural gas;
- general economic conditions;
- political considerations and government regulations/policies regarding exploration and development of a country's oil and gas reserves;
- advances in exploration, development and production technology;
- and the financial status of oil and gas producers.

Rig count is an important business barometer representing the activity level of the drilling industry, and this has an economic impact on most parts of the service sector. A 'rig' is a collective term to describe the permanent equipment needed for drilling a well (The Royal Bank of Scotland, 1996). It has come to include onshore and offshore vehicles, mobile platforms, or vessels on which the equipment is installed (i.e. jack-ups, semi-submersibles, drill ships). A rotary rig rotates the drill pipe from the surface to drill a new well (or sidetrack an existing one) to explore for developing and producing oil or natural gas.

The definition of rig counts varies amongst different companies. Baker Hughes (2002) defines that, to be counted as active, an international rig must be drilling at least 15 days during the month. A rig is considered to be drilling if it is turning to the right (i.e. the well is under way but has not reached the target depth or TD). Rigs that are in transit from one location to another, rigging up, drilling less than 15 days or are being used in non-drilling activities including production testing, completion and workovers are not included in the active rig count.

Other companies' approach to rig counts is different from that of Baker Hughes. Schlumberger's rig counts (2009) include rigs that are available or contracted but not actively drilling. These counts provide a census of rigs available for work rather than rigs that are actually working.

The M-I SWACO Worldwide Rig Count (2009) refers to a rig which falls into one of the following categories:

- it is engaged in drilling and related operations on the day of the count;
- or it is in transit to a location where it will commence drilling and related operations immediately upon arrival;
- or it has performed drilling and related operations for more than 15 days of the previous month.

Drilling-related operations include drilling, logging, cementing, coring, well testing, fishing, waiting on weather, running casing and blowout preventer (BOP) testing. The M-I SWACO Worldwide Rig Count may include completion activities if the rig continues directly from drilling and related operations to completion operations after reaching TD. Workover activities are explicitly excluded from the Worldwide Rig Count.

The drilling industry and many service suppliers follow the Baker Hughes International Rotary Rig Count, a monthly census of active drilling rigs for exploring or developing oil or natural gas outside the USA and Canada, as closely as the Dow Jones stock average.

According to Baker Hughes (2002), the active rig count acts as a leading indicator of demand for products used in drilling, completing, producing and processing hydrocarbons. When drilling or workover rigs are employed, they consume products and services provided by the oil service industry.

Changes in rig counts are associated with local government policies and/or political status, development of new infrastructure (such as roads and pipelines) and availability of capital investment. If these regulatory environmental conditions are constrained or unfavourable, the rig count drops (Noble Corporation, 2009).

Technology developments like minimising the number of wells required to develop a reservoir, maximising production from new and existing fields, increasing the operational efficiency of the active drilling fleet and opening new frontiers for exploration (such as deepwater areas) can also influence drilling activity and rig counts.

There are many other influences such as weather and seasonal spending patterns. Hurricanes can affect rig counts by forcing the evacuation of personnel from offshore platforms and delaying rig moves to new locations.

The following section will discuss major factors which may have a direct impact on oil and gas service activities.

2.4   Service-related petroleum economy

2.4.1   Oil price volatility versus steady offshore rig counts

Crude oil is priced internationally. Whatever the currency cost of extracting crude oil – and it is obviously much less expensive to extract from a desert in Arabia than from the depths of the North Sea – its revenue value to the operator is conventionally expressed in terms of US dollars per barrel. The USD price level and the USD exchange rate against national currencies both fluctuate and this volatility exerts a considerable measure of influence not only on the forward planning which all petroleum companies carry out, but also on any current E&P.

Gourlay (1996) discovered that, for many decades, these fluctuations were so small that they could be virtually ignored, but by the early 1970s there was soon to be dramatic change in this comfortable economic climate. Two unrelated but contemporary events, the collapse of the international currency system and political upheaval in the Middle East, brought about this change.

The world oil price has been especially dynamic since the 1970s (Figure 2.4) as a result of being driven by a number of sophisticated influences. In 1969, Qaddafi demanded an immediate increase in the posted price of Zelten (i.e. Libyan) oil, which was valued highly on account of its low sulphur content, and he successfully encouraged the other OPEC countries to demand similar increases. By September 1973, the price paid to the oil companies had risen to USD 2.90 per barrel from USD 2.48 in 1972. Then in October 1973 another Arab–Israeli war erupted. The Arab nations cut oil supplies by 25 per cent and this pushed up the price which by January 1974 had risen to USD 11.65 per barrel and then to over USD 22 on the wholesale market in Rotterdam.

Oil prices (USD per barrel)

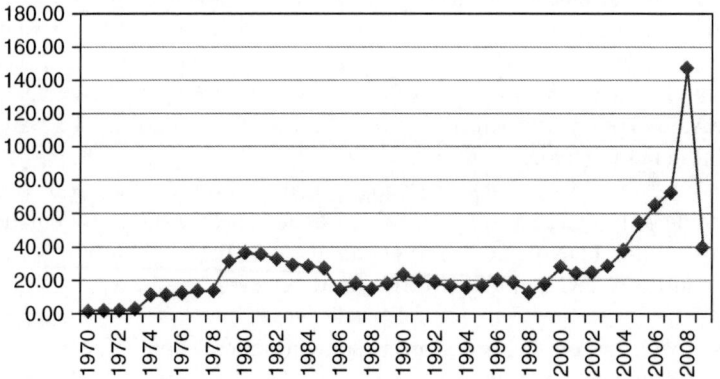

Figure 2.4   Global oil prices
Source: BP Statistical Review of World Energy (2009).

The price eventually stabilised at around USD 13 but this was a figure more than five times greater than a few months previously. As a result, the world experienced its first energy crisis (Gourlay, 1996).

The price of oil seemed to be on a permanent upward spiral which rose even more steeply in 1979 when the events of 1973 were repeated. OPEC raised the price of its oil by 5 per cent just before a revolution in Iran saw the ruling Shah ejected by a religious fundamentalist regime. With Iran in turmoil its output fell first by 6 million barrels per day, and then in September a further 4 million barrels per day vanished from the world's daily supply when Iraq attacked Iran. The price of oil doubled to USD 31.6. There were further OPEC-instigated rises of USD 7 in 1980 and USD 4 in 1981 and thus within eight years the price of a barrel of oil had jumped from less than USD 3 to USD 37 at one point (Gourlay, 1996).

A historical low point was reached in November 1998 of USD 10 per barrel, after increased oil production from Iraq coincided with the Asian financial crisis, which reduced demand. Prices then increased rapidly, more than doubling by October 2000 to USD 36, then fell until the end of 2001 before steadily increasing, reaching USD 40–50 by September 2004 (EIA, 2009).

Due to strong demand for petrol and diesel and concerns about refiners' ability to keep up, crude oil prices surged to a high of above USD 60 in July 2005. Prices rose steadily throughout 2006–7, from an average of over USD 65 per barrel in 2006 to an average of over USD 72 in 2007 (EIA, 2009; BP, 2009).

A mix of factors pushed oil prices to this level. OPEC proposed an output increase lower than expected. US stocks fell lower than experts predicted. There were changes in federal oil policies, and six pipelines were attacked by a leftist group in Mexico. Tensions in eastern Turkey and Nigeria, and decline of the US dollar, were also influencing factors.

In 2008, on the back of strong global economic expansion, particularly in China and India, there was sustained growth in demand for oil (PETRONAS, 2009). Over this period, there were a number of significant developments in the world petroleum industry:

- recurring supply disruptions in some producing countries and continuing geopolitical uncertainties particularly in the Middle East,

- global demand surpassed supply and growing concerns over security of supply,
- industry-wide speculative activities.

All these developments drove crude oil prices to historic highs in the first half of 2008.

According to EIA (2009), oil started 2008 with a single trade at USD 100, broke through USD 110 on 12 March 2008 and then roared to a record peak of USD 147.3 on 11 July 2008. After mid-2008, the trend reversed as global equities crashed. On 23 December 2008, oil fell to USD 30.3 a barrel, less than one-quarter of the peak price reached four months earlier. Hit by the global financial credit crunch and forecast for further reduction in world demand, oil prices did not rebound once 2009 started. Oil has since been trading between USD 35 and 50 a barrel in the first quarter of 2009 (EIA, 2009).

The fluctuation of the oil price directly affects the status of the service industry's market: when oil prices fall, activity slides. For example, a number of rigs in Europe and the US cold-stacked in the late 1990s as the oil price fell and stayed out of work through to around 2003 (Gourlay, 1996). In general, higher oil and natural gas prices or oil and gas producers' expectations of higher prices result in a greater demand for drilling services. However, sometimes higher oil prices do not necessarily translate into increased drilling activity since the oil clients' expectations of future commodity prices typically drive demand for rigs. That was the situation prevailing during the second quarter in 2009 when oil prices started to recover, but a number of international companies' (Ensco, 2009; Pride International, 2009) rig utilisation rates were lower than the corresponding period of one year previously.

Nevertheless, for the domestic-consumption-dominated economies like China and Malaysia, the dramatic adjustment in oil prices does not seem to have had a significant impact on the level of oil and gas E&P activities (Figure 2.5).

In contrast to the global economic slowdown, offshore oil and gas exploration activities in China have intensified. A cluster of discoveries in recent years has highlighted the potential of Bohai Bay. Since 2000, CNOOC has discovered a number of oilfields (e.g. Penglai 19-3, Caofeidian 11-1, Bozhong 29-4 and Bozhong 25-1) in Bohai Bay.

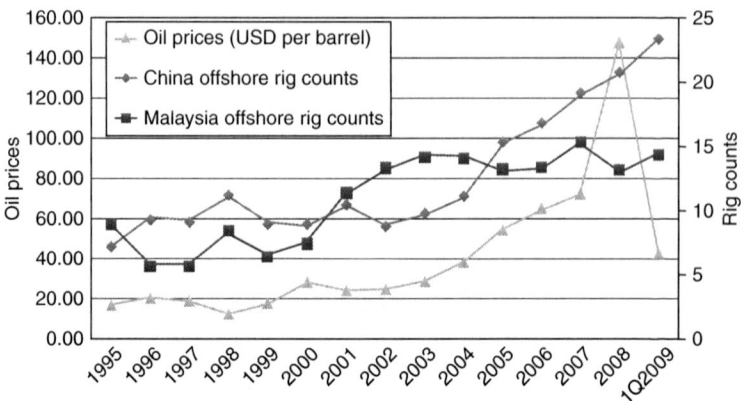

Figure 2.5   Sharp correction of oil prices versus steady trend of offshore rig counts in China and Malaysia
Source: BP Statistical Review of World Energy (2009); Baker Hughes (2009).

In March 2009, CNOOC announced two new oil and gas discoveries – Bozhong (BZ) 2-1 and Qinhuangdao (QHD) 29-2.

As can be seen from Figure 2.5, China's average annual offshore rig counts have grown consistently, from just 7 in 1995 to 21 in 2008. The highest rig count was 24, recorded in August 2008, after the worldwide credit crisis had started. The lowest offshore rig count of 4 was recorded in August 1999 (Baker Hughes, 2009).

Malaysia's rig counts grew from 9 in 1995 to 14 in 2008. A low rig activity was registered in 1999 with only 4–9 rig counts per month. The low activity shows an adverse impact of the 1997–98 financial crisis in Southeast Asia. The most active rig drilling in Malaysia reached a peak of 19 in December 2008. Overall, its rig counts have fluctuated in line with a level of just above or under 10.

There was a slow turnaround trend for rigs in both countries during the 1998–2000 Asian economic crisis and recovery period. In 1999, average rig counts were at a low level of 9 in China and only 6 in Malaysia. Although there was a sharp correction of oil prices, the drop has not been transferred to a declining trend in drilling activity. From 1995 to 2009, despite extremely volatile oil prices, offshore rig counts have shown a steady trend and have grown over the period.

2.4.2   Trends of petroleum economics

China is East Asia's largest oil player and produced more than 160 million tonnes of crude in 1999, making it the world's fifth-largest oil producer for 13 consecutive years. Oil output increased to 162.6 million tonnes in 2000.

As can be seen from Figure 2.6, despite the Asian financial crisis, China's oil production continued to grow. Onshore exploration and development activities in the western and the mature eastern regions have been enhanced. Annual oil output increased by nearly 26 million tonnes from 1997, reaching 186.7 million tonnes in 2007. Oil production rose nearly 2 per cent to 189.7 million tonnes in 2008 (BP, 2009).

Historically, natural gas has not been a major fuel but, given China's significant domestic reserves and the environmental benefits of using this energy source, a major expansion of gas infrastructure is taking place. Production has been rising steadily over the last two decades. It reached 30.3 billion cubic metres (bcm) or 27.3 million tonnes oil equivalent in 2001, up 30 per cent from 22.3 bcm or 21 million tonnes oil equivalent in 1998 (EIA, 2001; BP, 2001). Gas production has continued to grow. In 2007, China recorded the world's second-largest increment in natural gas production, up

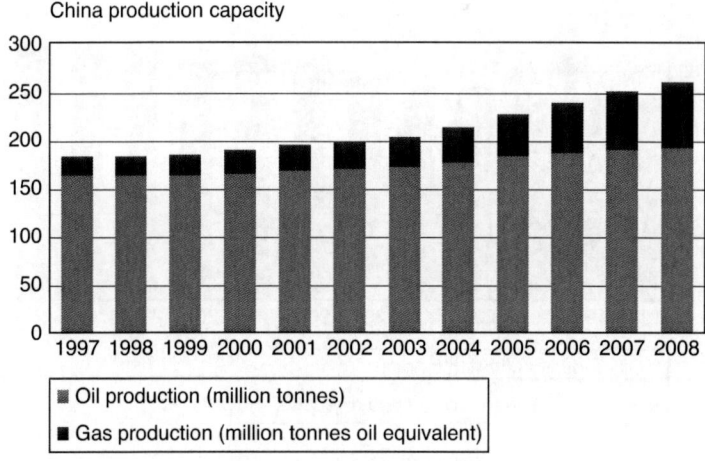

Figure 2.6   Oil and gas production in China, 1997–2008
Source: BP Statistical Review of World Energy, 2009.

18.4 per cent year-on-year to 69.2 bcm or 62.4 million tonnes oil equivalent. It ended 2008 with a 9.6 per cent growth year-on-year, reaching 76.1 bcm or 68.5 million tonnes oil equivalent for the year (BP, 2009).

As shown in Figure 2.7, Malaysia's crude production has been stable of date, with annual output fluctuating between 32 and 37 million tonnes over the period 1997–2008. In 2008, annual oil production was 34.3 million tonnes, as compared to a record high of 36.5 million tonnes in 2004. Over the period between 1997 and 2008, the lowest level of oil output was 32.9 million tonnes recorded in 2001.

In contrast, gas production in Malaysia has been rising steadily in recent years, reaching 47.4 bcm or 42.2 million tonnes oil equivalent in 2001, up from 38.5 bcm or 34.6 million tonnes oil equivalent in 1998. Thanks to its newly discovered deepwater gas resource, production rose nearly 30 per cent from 43.5 million tonnes oil equivalent in 2002 to 56.3 million tonnes oil equivalent in 2008.

China and Singapore own most of the mobile rigs and support vessels in the Asia-Pacific. It should be noted that Singapore's success in this regard and as a major service centre is despite this state having no oil or gas resources.

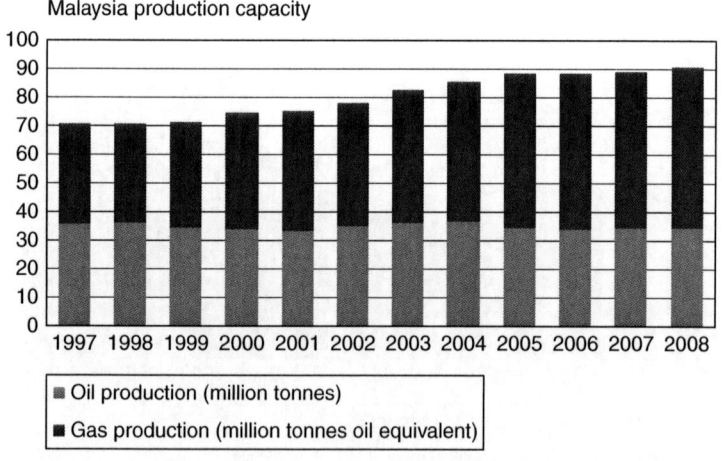

Figure 2.7   Oil and gas production in Malaysia, 1997–2008
Source: BP Statistical Review of World Energy, 2009.

While Malaysia is the largest offshore oil producer, China is achieving the highest output growth rates. Most Chinese oil production capacity, approximately 90 per cent, is located onshore. According to CNOOC (2009), China produced 158.5 million tonnes of oil in 1996. Of this, offshore oil production was estimated to total 14.6 million tonnes in 1996, some 9.2 per cent of total Chinese oil production. Offshore gas production in China was 1.3 bcm in 1996, which was 5.9 per cent of total Chinese gas production.

Chinese offshore production has more than doubled since 1997, which is ahead of the worldwide average. At 15 million tonnes, China's offshore oil production was nearly 10 per cent of its total oil output in 1999.

China's offshore oil production totalled 124.5 million tonnes in 2007, which was 12.9 per cent of the country's total oil output. Offshore total gas production was 91.2 bcm, which was 17.6 per cent of China's total gas output (CNOOC, 2009).

The above figures clearly illustrate the importance of the offshore oil and gas industry to China.

2.5   Political/regulatory factors

Because of economic, environmental or other policy issues, there are laws and regulations that govern exploration and development activities (Ensco, 2009; Pride, 2009). As a result, service activities are directly or indirectly affected by the petroleum-related fiscal and regulatory regimes of the countries where they operate.

In countries such as China and Malaysia, governments have regulatory restrictions aimed at protecting indigenous enterprises. There are formal or indeed informal requirements that local staff must form a certain proportion of the total. There is also an attention to 'localisation' whereby local companies are favoured.

2.5.1   China petroleum regulatory environment

In China, all industry regulations stem from government. The petroleum industry plays a pivotal role in its economy and receives considerable official support. The industry will continue to enjoy this status as Chinese leaders have quickly grasped the essentials of energy-security issues in an import-dependent environment (China–Britain Business Council Scotland, 2000).

Beijing has advanced four basic strategies for its petroleum industry: maximum development of domestic resources, creation of strategic reserves, seeking foreign technology and investment, and making strategic investments in upstream production abroad (IEA, 2000; Chen, 2000).

With regard to the offshore oil and gas industry, a regulation on foreign involvement in exploitation of offshore petroleum resources was issued by the State Council early in 1982. Under this rule, only CNOOC has the monopoly right to carry out offshore exploration, development and production activities in the foreign cooperation sea water area.

Any foreign oil operators who wish to explore in Chinese waters must sign contracts with CNOOC. During the exploration phase, foreign operators have to bear exploration costs and similarly costs of development of any commercial finds. CNOOC has the right to own, unconditionally, 51 per cent of any resources discovered by such foreign parties.

Despite the fact that CNPC and Sinopec hold a number of offshore oilfield blocks, they do not have the right to take part in foreign cooperation. Unlike CNOOC, the two oil giants cannot benefit from foreign financial resources and technological advances through foreign cooperation. The two state-owned companies have been lobbying the administration department constantly.

Efforts made by CNPC and Sinopec do, however, seem to be paying off. In March 2009, it emerged that CNOOC will lose its offshore monopoly rights in the sphere of foreign cooperation. CNPC and Sinopec will be allowed to carry out exploration, development and production activities with foreign partners.

Domestic policy encourages Chinese oil companies to use Chinese-owned drilling rigs at a lower rig day rate than the international standard. Most of the E&P companies are either totally or partially owned by the Chinese. By applying location and policy advantages, these Chinese players are able to keep a hold on most domestic drilling spending, leaving little for industry followers or outsiders.

2.5.2   Malaysia petroleum regulatory environment

The Malaysian government polices allow indigenous firms to enjoy more benefits than other inward-investing companies. The government conducts a protectionist policy by regulating strict requirements on investment. Like China, there are initiatives that promote indigenous suppliers. All companies supporting the upstream sector must be licensed by PETRONAS. Those supplying the downstream sector must also be registered.

Companies wishing to supply equipment, services or materials to the upstream sector must partner a local firm run by Bumiputras (*bumiputra* means 'son of the soil' and refers to the indigenous Malays). The *bumiputra* policy is designed to encourage participation of indigenous Malays in the Malaysian economy which, at the time, was dominated by Malaysian Chinese, and foreign business people (DTI, 1996). The Vendor Development Programme is a particular initiative that ensures locally manufactured goods are used in preference to international goods; only high-tech equipment not available locally is exempt from this programme (Britain Organisation, 2001).

2.5.3   Singapore petroleum regulatory environment

The Singaporean government has set up very supportive policies towards the oil and petroleum industry. Appreciating its strategic location and attractiveness to foreign investors, Singapore established itself as the region's leading oil refining and petrochemicals centre long ago. It has become the number one logistics support basis for upstream activities in the region since it began its industrialisation programme in the 1960s (DTI, 1996). Because of the high value-added nature of the petroleum industry, Singapore targeted this sector to complement its port facilities. Over the years, the government worked closely with the industry to make sure that its strategic planning development kept pace with integrating processes (The USA Embassy Singapore, 1997).

2.6   Technology factors

The oil and gas industry's future success is seen to be linked to technology. Technological advances could help to lessen exploration costs by reducing drilling and production costs, and therefore increase profit margins. New environmental protection techniques have been generated in an attempt to reduce potential risks or increase the level of protection. What is more, advanced technologies are increasingly important in achieving and maintaining a competitive advantage (Pearce and Robinson, 1997).

Petroleum technology development has been driven by a number of factors. Overall, more technology is needed to maximise oil and gas production. At the end of 2008, approximately 15 billion barrels of oil remain in known reservoirs in China, and 5 billion barrels in Malaysia (BP, 2009). These resources could potentially be recovered by advanced technology.

According to Ellix (2002), major trends in petroleum technology development include but are not limited to:

- Squeezing the last drop from older fields
- Managing old platforms and pipelines
- Unlocking small fields
- Challenges of heavy oil and HPHT (high-pressure, high-temperature) fields
- More accurate and successful exploration
- Deepwater/ultra-deepwater subsea technology
- Maintaining safety and environment

Advancing petroleum technology has become of greater importance as offshore exploration increasingly focuses on deeper water, Arctic and other frontier areas. PETRONAS (2009) says that, without the previous development of advanced technology, production in these areas would not be possible.

Technology need tends to be associated with regional characteristics. Many unexplored areas in Asia may be in harsh (*taifeng* – hurricane) or sensitive (rainforest) environments, where current known techniques may not provide sufficient solutions.

In general, only the largest companies in the industry possess the technological sophistication and financial resources to apply advanced technologies successfully (Pearce and Robinson, 1997). Application of advanced technologies by upstream oil and gas companies has yielded strategically valuable outcomes. They can better define drilling prospects, and thereby lower the risk of a dry hole by, for example, applying 3-D seismic technology. They can drill a well faster or less expensively with improved drilling 'mud' and measurement-while-drilling technology. It is also possible for them to increase the volume of oil produced from a reservoir through, for example, fracturing technologies, horizontal drilling and other enhanced oil recovery techniques.

Nevertheless, even though technological advances create competitive advantages, they tend not to be well diffused throughout the industry or across international regions. Only the largest oil companies within the industry engage in E&P-related R&D. Even with old technologies that the major companies no longer hold as proprietary, technology transfer to smaller companies has been poor (Pearce and Robinson, 1997).

Service companies that master the application of such advanced technologies can be more successful than their competitors that lack similar capabilities. In particular, expertise in geology, drilling processes, and oil and gas recovery technologies brings high added value since it can help oil clients to exploit opportunities more successfully than other companies.

The Chinese industry has to a great extent been self-sufficient, with its own geological, geophysical, drilling and production facilities. However, specialist technologies have been imported from the West since the early 1980s (Scottish Enterprise, 2002). Although China produces equipment for onshore exploitation, it is estimated that about 80 per cent of the equipment needed for offshore activity has been imported (STI, 1995).

Some Chinese enterprises, such as Chiwan Sembawang Engineering Company Limited (CSE), belong to the labour-intensive side of manufacturing. Companies such as CITIC Offshore Helicopter Company Limited (COHC) cooperate with international players so that they can use advanced foreign experience for reference.

Malaysia also has to import advanced technologies from abroad. In this regard, foreign-owned service companies are more competitive than domestic small and medium-sized enterprises (SMEs) that do not have the same advanced technologies.

Quite often, indigenous companies' access to new or advanced technologies has not been easy not only because of competition issues but also on account of the policies of foreign companies' governments. Some Western countries operate restrictions that have prevented the export of certain advanced technologies to countries like China.

2.7   Clients of oil and gas services

2.7.1   National oil companies (NOCs)

China's oil and gas industry is monopolised by the three key state-owned enterprises (SOEs). CNPC, Sinopec and CNOOC are controlled by China's State Council. The SETC coordinates the majority of economic activities within the oil and gas sector and supervises the operations of all subsidiary organisations linked with these three companies.

CNPC produces crude oil onshore for its domestic refineries as well as supplying those of Sinopec. Sinopec's major business is refining, but with a few oilfields, it also produces crude oil onshore and offshore.

Over the past 40 years, CNPC has grown into a multidisciplinary vertically integrated organisation. It has engaged not only in petroleum exploitation and processing, but also in the manufacture of relevant equipment as well as in other diversified businesses.

CNPC's own factories produce 60 per cent of the equipment which it uses in E&P (STI, 1995). By the late 1980s, the company was divided into more than 20 divisions, with each being responsible for a specific geographical area (Scottish Enterprise, 2002).

While constantly improving its shallow-water installations, CNPC is exploring opportunities in deepwater areas and other harsh marine environments both domestically and abroad. CNPC has extensively integrated service capacities. As of end 2007, CNPC owned 1044 drilling rigs. In support of its offshore exploration and development, CNPC carried out construction projects including the artificial island in the Jidong field and the Qindao Offshore Engineering Construction Base (CNPC, 2009).

PetroChina, a CNPC subsidiary, owns 13 large oilfields: Daqing, Jilin, Liaohe, Huabei, Dagang, Jidong, Changqing, Sichuan, Xinjiang (Karamay), Tarim, Tuha, Yumen and Qinghai. In 2000, it produced around 66 per cent of China's total output of oil and gas.

CNOOC focuses on domestic offshore E&P, frequently in cooperation with foreign oil companies. It also owns the China Offshore

Oil Research Centre, a chemicals company, eight specialist service companies and five logistics companies (CNOOC, 2009).

CNOOC claims to own a wide range of offshore rigs, engineering vessels and other special equipment, most of it conforming to world standards. CNOOC's equipment is not only used by itself, but also rented by foreign companies (STI, 1995).

PetroChina, Sinopec and CNOOC have significant experience of developing partnerships with foreign companies, both in terms of joint ventures in exploration, development, petrochemicals and downstream activities, and of being significant purchasers of foreign plant, equipment, technologies and services. These three companies continue to have the state as their major shareholder and remained regulated by the SETC after they were listed on the New York, London and Hong Kong stock markets in 2000. The three started venturing overseas, seeking E&P opportunities, mainly to North Africa, South America and former Soviet countries (Scottish Enterprise, 2002).

An emerging trend shows that Chinese state-owned oil firms have increasingly acquired interests in the Central Asian region, the Middle East, North Africa and South America. The focus now is to expand these interests to Russia, Kazakhstan, Turkmenistan, Iran, Iraq, Sudan, Venezuela and Indonesia.

Compared to China, the components of national oil companies in Malaysia and Singapore are less sophisticated. PETRONAS is an active driller and operator, being engaged in almost every sphere of activity in the Malaysian petroleum industry, including the oil and gas service sector. Its corporate structure comprises a number of wholly owned subsidiaries and other partially owned or associated firms (DTI, 1996).

Through PSCs at home and abroad, PETRONAS is actively engaged in the exploration, development and production of oil and gas. To augment domestic petroleum resources, PETRONAS's overseas operating arm, PETRONAS Carigali Sdn Bhd, has since 1990 undertaken exploration, development and production activities outside the country. Currently the company is actively involved in the upstream sector in 25 countries (PETRONAS, 2009).

In Singapore the national oil company, the Singapore Petroleum Company (SPC), is a major indigenous upstream player (PetroMin, 2000). Founded in 1969, SPC has grown from a predominantly home-based company to a regional player with a presence in Asia. Today, SPC is an integrated petroleum operator whose business activities cover oil and gas E&P, gas pipelines, as well as other downstream activities (SPC, 2009).

Historically, state-owned companies in China and Malaysia have been the key buyers of the oil and gas service sectors. However, in recent years, with more and more international oil firms stepping into E&P activities in the domestic markets, the buyer structure has been widened. Buyers from international companies, emerging joint ventures and the wider energy or marine industries are adding to the service companies' customer profiles.

2.7.2   Multinational oil companies

Major international oil firms (IOCs) with production interests in China include big names from North America, Europe, Southeast Asia and the Middle East. The number of international operators is still on the rise.

Offshore, CNOOC has signed a number of contracts and agreements with various international companies, including BP, Shell, Kerr-McGee and Global Santa Fe (Scottish Enterprise, 2002). Other international buyers for offshore services in China include companies such as ConocoPhillips, Chevron, CACT-OG, Japan China Oil Development Corporation, and Hyundai Heavy Industries.

Offshore oil and gas projects can be divided into two categories in China. That is: joint exploitation between CNOOC and foreign companies; and CNOOC self-financed exploitation. It is essential to identify the designers and owners of the projects concerned when seeking to supply equipment. Joint exploration projects through production sharing agreements (PSAs) are always conducted by a foreign company that begins exploration at its own risk, having successfully bid for the contract. In this scenario, suppliers would be expected to approach the foreign company rather than the Chinese side.

In Singapore, there are many major international oil and gas operators (PetroMin, 2000). For Malaysia, by 1996 the only international operators offshore were Esso and Shell (DTI, 1996). Since 2000, international producers have increased dramatically and include:

- Amerada Hess
- BP
- Eni
- ConocoPhillips Sabah Gas Ltd
- Malaysia–Thailand Joint Authority (MTJDA)

- ExxonMobil
- Murphy Oil Corporation
- Newfield Exploration Company
- Pertamina
- PetroVietnam

- Santa Fe Energy Resources
- Shell Energy Asia Limited
- Talisman Malaysia
- YPF Malaysia

They all have been awarded PSCs to carry out exploration, development and production activities in Malaysia.

2.8   Service suppliers

Service companies range from small to very large indigenous and international firms, including technology- and labour-intensive enterprises. Leading suppliers vary in each of niche service segments and they come from different countries (STI, 1995; DTI, 1996). For example:

- Mobile offshore drilling units (jack-up or semi-submersible), are provided by suppliers from Japan, Singapore, Norway, Sweden and China;
- Support vessels are mostly supplied by companies from Japan, Norway, Malta, Singapore, Denmark and China;
- Contractors and consultants come mainly from the US, Japan and Korea; such companies tend to have a strong foothold in the region;
- Maintenance and repair work and related supplies tend to be handled by local companies;
- International companies are dominant in the inspection sector with few domestic firms having the capacity to carry out such work.

In general, small companies are not well known and are difficult to identify because they go in and out of business, alter their corporate identities and change addresses frequently (Pearce and Smith, 1997). Some world-leading giants such as Schlumberger, Bechtel, Halliburton and Baker Hughes operate widely in East Asia, but numerically they comprise only a small portion of all service companies in the region. Consequently, it is difficult to determine accurately the number of energy service companies active in East Asia.

2.8.1   National service companies

In China, service firms used to be owned by the three state-owned companies. Now the ownership structure is changing and many of these companies have been privatised as a result of Chinese industrial restructuring.

CNPC, Sinopec and CNOOC have numerous subsidiaries or institutions of their own to support and service every activity from exploration to production and refining. For instance, the China Offshore Oil Development and Engineering Corporation (COODEC) was, in the 1990s, the only design institute for China's large offshore projects. CNOOC self-financed projects were designed by this organisation (STI, 1995).

Today, China Oilfield Services Limited (COSL) and China Offshore Oil Engineering Company Limited (COOEC) are the two major offshore service affiliates of CNOOC. CNPC's major offshore service arm is one of its subsidiaries called the China Petroleum Offshore Engineering Company (CPOE).

COSL is a provider of integrated oilfield services specialising in drilling in offshore China. Its core service businesses cover each phase of oil and gas exploration, development and production, including drilling services, well services, marine support, and transportation and geophysical services (COSL, 2009). Today the company is active globally.

Headquartered in Tianjin, COOEC is the largest offshore engineering and construction company in China. It provides integrated services through engineering design, onshore fabrication, offshore installation and maintenance, underwater engineering, and alternative energy projects. COOEC is one of the largest offshore engineering, procurement, installation and construction (EPIC) contractors in the Asia-Pacific. It has carried out operations in offshore China and in the Middle East, Southeast Asia and South Korea.

As a strategically important offshore business unit, CPOE has total support from its parent company CNPC in policy, finance and capital, human resources, technology and marketing. It was formed as a joint venture by CNPC, Liaohe Petroleum Exploration Bureau and Dagang Oilfield Group in November 2004. CPOE is mainly engaged in offshore engineering and technology services, which have been predominantly deployed in the shallow-water areas of Bohai Bay.

Other big, specialised, and publicly listed service organisations include the CITIC Offshore Helicopter Company (COHC), providing offshore helicopter services, China State Shipbuilding Corporation (CSSC) and China Shipbuilding Industry Corporation (CSIC). Both CSSC and CSIC specialise in FPSO construction and offshore engineering equipment.

In Malaysia, PETRONAS has played a pivotal role in the oil and gas service industry. PETRONAS Research & Scientific Services Sdn Bhd is a subsidiary which undertakes much of the research and development activities for its parent. Another subsidiary, OGP Technical Services Sdn Bhd, provides services ranging from basic design, engineering and procurement to construction management for the oil, gas and petrochemical industries domestically and internationally.

There are other significant service companies such as Malaysian Energy and Maritime Services Sdn Bhd, Malaysia Marine and Heavy Engineering (MMHE) and Dialog Group Berhad. MMHE's three core businesses comprise engineering and construction, marine repairs, and marine conversion. It fabricated the truss spar floating production unit (FPU) and FPSO for the Kikeh Development Project, Malaysia's first deepwater development.

Singapore is the top player in the global market for mobile offshore drilling units and offshore support vessels. According to Singapore EDB (2009), there is a high probability that a modern jack-up rig working just about anywhere was designed or manufactured by a Singapore-based company. Local conglomerates Keppel and Sembcorp Marine are internationally renowned names in the global oil and gas industry.

Sembcorp Marine is especially prominent for its rig building and offshore conversion expertise. It also specialises in ship repair, shipbuilding, ship conversion, offshore engineering and construction. With its engineering, procurement and construction (EPC) capabilities, the company has established a track record in jack-up design and fabrication, semi-submersible construction, offshore platform production, and the conversion of FPSOs (Sembcorp, 2009).

2.8.2   International service companies

The Asia-Pacific has become the fastest-growing regional economy in the world and the need for energy is expanding as well. It is in this context that many international companies are pursuing large opportunities in the region's rapidly developing petroleum market.

With highly skilled workforces and access to emerging economies, nations such as China and Singapore have become a logical choice as the regional headquarters for international energy service companies. Many international companies have established their own logistic hubs in Southeast Asia, allowing easy access to opportunities and resources.

In China, international oil engineering firms such as Halliburton, Baker Hughes, Schlumberger and Technip actively collaborate with their Chinese oil clients and enjoy strong traction in the local market for their advanced technologies, instruments, tools and equipment (Scottish Enterprise, 2002).

International service contractors also have to work closely with local companies in Malaysia. Ipedex (SEA) Sdn Bhd, for instance, is part of the International Group of Ipedex Companies and it was incorporated in Malaysia in 1992 in order to develop its activities in South and East Asia (Ipedex, 2007). Other internationals such as Fluor set up offices in Malaysia to enhance their presence in the regional market (Fluor, 2007).

Since the 1960s, Singapore has developed local infrastructure and expertise for the production of both up-hole (above ground) and down-hole (below ground) oil and gas equipment. The equipment produced in Singapore has been supplied to major oilfields throughout the world. Suppliers of oil and gas equipment comprise more than 100 companies including 60 international companies and local SMEs. Global service giants, such as Schlumberger, Halliburton, FMC, Cooper Cameron and Baker Hughes, have a major regional presence there (Ng, 2009). Today, Singapore has also become the front-end engineering and design (FEED) centre for the Asia-Pacific region. Numerous specialist international dealers service the larger companies, providing opportunities for small or medium-sized overseas firms to supply equipment or components.

2.9   Competition in the service sector

2.9.1   A framework of industry rivalry

In general, energy services are very competitive and cyclical. There have been times of high demand and short supply when service companies charge premium prices, followed by times of lower demand, oversupply and service companies having to cut prices sharply (PHI, 2009). Over the period 2002 to mid-2008, for instance, the industry was on a cycle of high demand, short supply and higher prices charged to clients.

Many types of service contract are awarded on a competitive bid basis in the selected three countries. Winning a contract with an operator is dependent on a number of key successful factors (KSFs) linked to competition:

- price or price package
- quality of products/services
- equipment availability and suitability
- equipment condition
- experience or track record
- safety performance record
- operating integrity
- existing positioning in the industry
- customer relationship
- location

Market conditions in the oil and gas service industry may change from time to time. Service companies may respond by deploying different competitive strategies (Figure 2.8).

In order to be competitive, higher-level capital investment is generally involved as service companies quite often undertake expansion and/or upgrades of their asset capacity and/or capability. The increase in supply by new entrants or new asset capacities and changes in existing suppliers' capacities may intensify price competition (Pride, 2009).

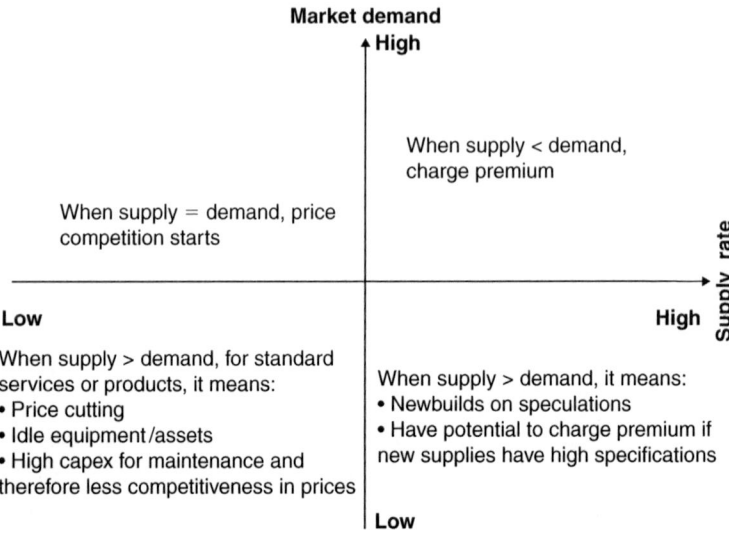

Figure 2.8   Rivalry, demand and supply in the energy service industry

When the service market is overheated by supply, some assets may drop out of the market. Oversupply of products and services will tend to depress prices. Carrying idle inventories for a long period could prove too expensive so service suppliers may in this case cut prices heavily to secure a contract (Ensco, 2009). Only high-specification products or services such as deepwater operating capacities offer premium pricing potential (Figure 2.8).

The consequence for industry profitability of an imbalance between supply and demand differs widely from segment to segment. Some service segments trigger price wars and low profitability even if there is little change in the amount of excess capacity. In oil tools and many other oilfield equipment/products, for example, there was an intense price cut during the 1997–99 oil price slump.

In other segments, periods of excess capacity have had relatively little impact on profitability because of the favourable service industry structure. In the Chinese offshore drilling market, during the period 1997–2000 for instance, there was relatively little discounting as the market was then dominated by only three Chinese suppliers, CNSPC, CONDC and COSDC.

2.9.2 Competitive forces

Entry threat

Barriers to entry into oil and gas services are high because it requires high capex requirements, a need to establish effective supplier–buyer relationships, and the need to ensure full and proper attention to safety. Many service firms are obviously always on the lookout for rivals, but existing suppliers are normally so entrenched in their area that any competitor would find it difficult to enter the arena and challenge them.

In some sectors such as drilling, FPU services, helicopter services, offshore supply vessels, oil tanker shipping and offshore pipe-laying, the existing barriers are very high because of the high specialisation required. In price terms, this has translated into 'short peaks and long troughs' (Bristow, 2009; SEACOR, 2009). When there are prolonged periods of excess capacity, such barriers keep firms from leaving oilfield services.

For international companies, entry barriers include local content issues. However, leading international service companies can overcome the 'localisation' barrier by providing products or technologies specific and focal to countries like China, Singapore and Malaysia. It is difficult for domestic companies to compete with these internationals, especially when it comes to deepwater technology and services.

Threat from substitutes

In general, substitutes are not a threat to oil and gas service contractors. In most cases, oil companies sign contracts for highly specialised services. Chinese and Malaysian oil companies have their own assets or equipment, or the capability to support operations offshore themselves.

Buyers' power

As already emphasised, in China and Malaysia, this industry is monopolised by state-owned companies because they have the power of authority. The service sector is in a weak position in any type of buyer–seller negotiations. As the number of oil clients is stable in China, with only three major buyers CNPC, CNOOC and Sinopec, competition is not influenced by fluctuation in the size of the client base.

With regard to multinational petroleum clients, they normally have strength in finance and technology, leaving suppliers very limited room in which to negotiate. Furthermore, mergers among international oil and natural gas E&P companies may from time to time reduce the number of available clients, resulting in increased price competition.

Pressure from downstream suppliers

Apart from competing for clients, service companies may also have to compete for resources. Shortages of lower-end suppliers and services, skilled people, financial resources and needed equipment have happened in the past and could repeat again in the future.

Convergence of demand globally has resulted in shortages of getting certain specialised technical staff. Especially in Singapore, there are manpower shortages and in that situation, attracting technical talent becomes a competitive issue. A ready supply of labour from China, the Philippines, India and to a lesser extent Indonesia is alleviating some of the strain in Southeast Asia.

China has been training hundreds of thousands of engineers and other qualified personnel. Some Chinese companies even have the capacity to compete with Western contractors, offering their services throughout the world. Additionally, Chinese service firms can gain support from their parent companies in technology, policy and availability of financial resources.

Foreign service companies may need to bring in their own technicians from elsewhere. They generally have a highly mobile workforce and easy access to equipment at their local logistic bases. International companies can provide specialised service, product knowledge and industry experience to regional clients anywhere in the world.

Turning to logistics, each country has its own bases. For example, Shengzhen Chiwan base and Bohai Bay are the two offshore logistics hubs in China. In Malaysia and Singapore, Kemaman Supply Base (KSB), Asian Supply Base, Loyang Base and Jurong Marine Base have secured the supply spaces for international and domestic contractors.

Industry rivalry

The East Asian market is very competitive and the crucial issues are pricing and after-sales service. Competition comes from domestic and international players that have a strong foothold in the selected countries. In many industry sectors, competition may be dominated by a few major and several small competitors. Most domestic companies compete locally, with larger ones also going international to compete on a regional or global basis. The number and supply capacity of service players have been changing in China, Singapore and Malaysia.

A recent trend that is influencing the industry rivalry is the consolidation in the service and supply industry. Big Chinese steel manufacturers, for instance, are merging with each other to expand production capacity, or acquire their strategic suppliers in order to enhance production efficiency (BAOSTEEL, 2009).

Another trend is that service suppliers have started to spin off non-core business assets. In 2009, China Petroleum Pipeline Bureau (CPP), a CNPC subsidiary, sold its gas assets so that it could focus on core onshore and offshore pipe-laying businesses (CNPC, 2009).

Large state-owned contractors such as CSSC have restructured in an attempt to establish a global brand. With the full support of its government, CSSC has become a remarkable top player in the world marine and offshore engineering industry.

With regard to the competition in the offshore services market, firms tend to have advantages over others when they display financial and technological strengths. Firms with greater financial resources to cover high capital and maintenance costs and/or, in particular, firms differentiating themselves with unique products or innovative technologies, may outperform their peer competitors.

As a result of competition, some service companies have withdrawn from the East Asian market. Some of these companies went out of business or were sold, and many others sold off their existing assets and new units under construction or changed their business focus. For example, Sembawang's business in China today is very different from ten years ago.

2.10 Summary

In spite of the Asian financial crisis in the late 1990s and the 2008–9 oil price volatility, the business environment in East Asia has been benign for oil and gas service activities. Compared with China, Singapore and Malaysia appear more vulnerable in terms of confronting regional and global economic crises.

Reform of the Chinese petroleum industry has created room for companies to have greater control of resources and more freedom to manage their own affairs. Consequently, the development of China's service sector has become more market-oriented. In Malaysia, all service companies are influenced by government policies supporting indigenous Malays. Thanks to its government support, Singapore has become a logistic hub serving the upstream oil and gas activities in Asia. Nonetheless, petroleum-related political/regulatory factors or localisation policies may be hostile towards service companies, especially internationals.

Petroleum technology developments in Asia are moving to off-shore exploration with a focus on deeper water and harsh or sensitive environments where the existing technologies may not provide sufficient solutions. Offshore technologies have not been diffused completely from IOCs to NOCs. China has been self-sufficient in onshore technologies but the majority of its offshore equipment has been imported. Malaysia has also imported advanced technologies from abroad. In this sense, domestic service suppliers in China and Malaysia are less capable of providing competitive advantages compared with international counterparts.

The oil and gas service industry in East Asia comprises both indigenous and international suppliers. Mergers and acquisitions, organisational restructuring, and changed ownership structures have substantially reduced the number of companies active in various niche industry sectors. All Chinese service firms were once wholly owned by the state. Now the ownership structure is changing through joint ventures (JVs), joint development, wholly foreign-owned and domestic ownership shifts. Consolidation among international suppliers has also altered the number of service firms in Asia.

Competition from domestic suppliers is symbolised by the movements of capacity expansion, globalisation and niche market concentration. Internationals compete in the three selected countries mainly based on their technological strengths.

Historically, state-owned companies in China and Malaysia have been the key buyers. Today more and more international oil firms are stepping into E&P activities in East Asia, and the buyer structure has widened.

3
Conceptual Frameworks of Business Environment and Strategies

This chapter develops a selected review of the literature on the business environment. It also deals with major thinking and research on strategy, which is written comparatively and chronologically from a different philosophical and historical perspective between the East and West. The variables of business environmental dimensions, business strategic characteristics, and overall indicators of strategic performance are summarised. It also develops theoretical frameworks drawn from the literature review in order to test the data gathered from the empirical investigation.

3.1 The conceptual frameworks

Three objectives are established regarding the development of conceptual frameworks. Firstly, it sets out a methodology for understanding the total environment in which a company operates, and identifying and analysing those significant elements or sectors. This chapter outlines some theoretical notions and a framework for thinking about the business environment. A conception of environment built around three levels is presented. An analytical model for engaging in environmental analysis is developed. The critical activities in environmental analysis – scanning and assessment – are described in detail.

The second objective is to review strategic theories so that a framework for the investigation of business strategies can be developed. It offers a historical perspective on how strategy definitions evolved from military use into a discipline within the field of strategic management. The chapter explores chronologically various schools of business strategists and the major thrust in each of their arguments. An overview of the major work that has been carried out in the area of strategy development in the past and an outline of how it will develop in the future are presented. The applications of strategies in both Western and Eastern contexts as well as in the oil and gas industry are included.

The third objective is that the conceptual framework brings together business environment models and strategy concepts. A typology of business strategies and a strategic performance assessment approach are also generated in this chapter.

3.1.1   A generic model for strategic management processes

A formal strategic management process can be progressed in five major phases when an organisation has operated businesses for a number of years (Figure 3.1). Phase one involves an evaluation of current strategies. In this stage, corporate, business and functional strategies should be clearly identified or clarified. In particular, business strategies are

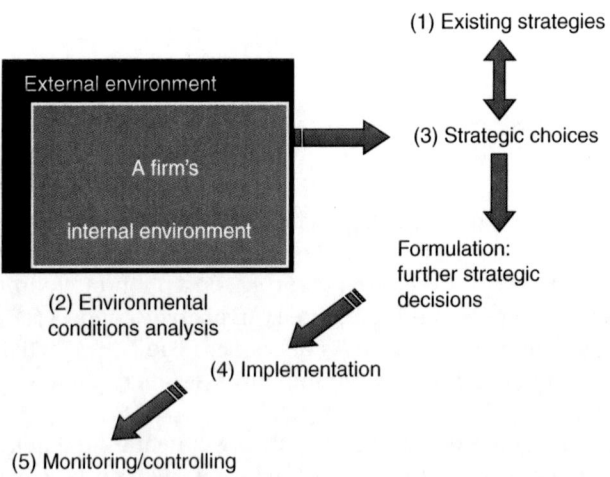

Figure 3.1   A generic model of the strategic management process
Source: Based on Pearce and Robinson (1997).

distinguished from business tactics – for example, niche focus competitive strategies versus marketing segmentation approaches.

Phase two is concerned with information gathering, strategic scenario analysis and new strategic options. Environmental (internal and external) scanning and assessment of progress are conducted at this stage.

Phase three pertains to developing models and frameworks for existing strategic business units. A business process may be remodelled and current strategies can be further developed. The above three phases form the process of strategic formulation.

Phase four is a strategic implementation process. The defined strategies and the strategic process should be introduced to the organisational management team and thereby they can be executed at all levels within an organisation. If necessary, appropriate training can be arranged.

In phase five, the ongoing process of the strategic management assessment is carried out to monitor the expected outcomes of the adopted strategic solutions. This may include the development of a strategic performance assessment system. This process can also be called strategic monitoring and controlling.

Within this conceptual context, the scope of the present research is limited to three focused aspects: the business environment assessment, the identification of existing strategies and evaluation of strategic performance. More details relating to these three aspects will be given in the following sections.

3.1.2 Business growth-related grand strategies

An oil and gas service organisation can conduct its operations at both a corporate level to achieve growth and a business level to compete in the marketplace. Its core businesses grow through both internal and external development.

For internal development or organic growth (Pearce and Robinson, 1997), there are five strategic approaches which have been adopted by energy service companies:

- Concentrated growth is to increase the use of current services provided by existing business units in the existing markets.
- Market development is concerned with finding new markets for the current services, i.e. seeking opportunities in the international markets or other upstream sectors.

- Service or product development involves finding new solutions to provide services.
- Innovation involves finding new technologies which can be applied by the company and its end users for the present markets or clients.
- Business 'fission' means that when a business unit becomes big, it can be split up into small parts for the purposes of effective management.

For external development, integration and diversification as well as cooperation strategies are essential strategic options. Pearce and Robinson (1997) define these external development strategies as shown in Figure 3.2:

- Integration includes horizontal and vertical integration. Horizontal integration seeks to pursue growth through the acquisition of one

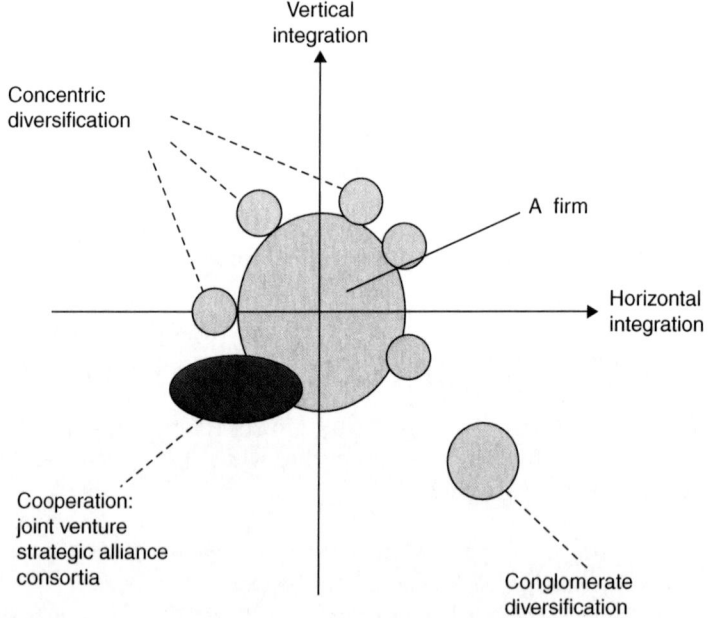

Figure 3.2   Corporate growth by external development
Source: Based on Pearce and Robinson's (1997) definition.

or more similar firms operating the same stage of the production–market chain. Vertical integration acquires firms that supply the company with inputs (raw materials) or are customers for its output (warehouses for finished products).

- Diversification includes concentric and conglomerate types. Concentric diversification refers to acquisition of businesses that are related to the firm in terms of technology, markets or products. Conglomerate diversification refers to acquisition of a business representing the most promising investment opportunity and it is based on profit considerations.

- Cooperation strategies include joint ventures, strategic alliances and consortia.

3.1.3   A model of environmental school analysis

Duncan (1972) defines a concept to distinguish the external environment from the internal. The internal environment refers to all factors existing inside an organisation, including organisational culture, structure and functions. The external environment refers to all the forces outside the organisation. In this study, the term 'environment' refers to the external environment.

Many well-known researchers (Brooks and Weatherston, 1997; Mintzberg et al., 1998; Weick, 1979) have categorised the different approaches to defining the environment. Strategic, organisational and international business scholars and researchers develop the conceptions of the business environment from three different perspectives, and suggest that the business environment may be viewed as the objective, perceived and enacted environment (Figure 3.3).

The objective environment (Bourgeois, 1980; Dill, 1958) is a clear, measurable and definable reality. It emphasises the external existing facts which affect an organisation. Nearly all strategic management literature includes this assumption that there are factors that all organisations face and deal with. The business environment can be understood only if it is described and measured.

The perceived environment 'remains real, material and external', but involves strategists' 'incomplete and imperfect perceptions of the "environment", and focuses on minimizing the gap between their flawed perceptions and reality of their environment' when they set up missions, goals and objectives (Mintzberg et al., 1999). This is

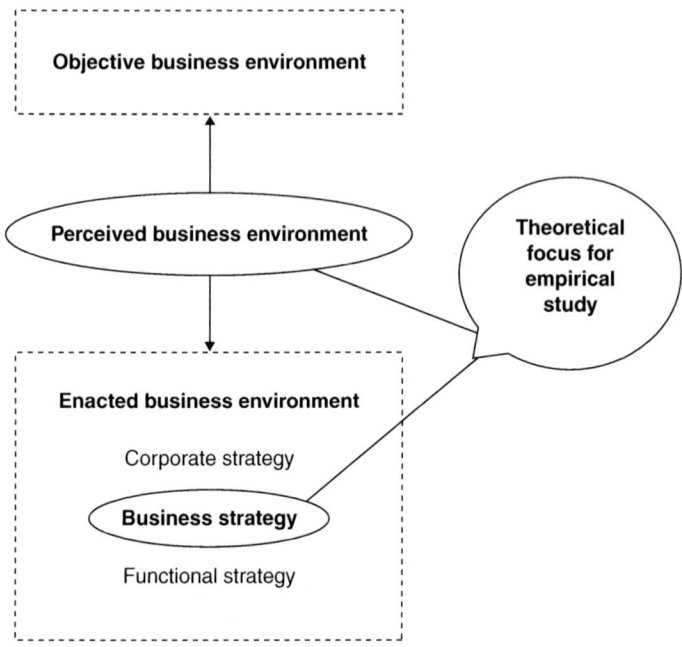

Figure 3.3   Focus of the empirical research

a subjective fact. The particular characteristics of the environment rely on each individual's interpretation and perceptions. Brooks and Weatherston (1997) point out that 'different organisations in the same industry often "view" environmental forces quite differently from one another, even though those forces may in fact be very similar'.

The enacted environment refers to companies' strategists linking the business environment to their behaviours by using strategies and developing strategic models or using existing models to test them (Brooks and Weatherston, 1997). Weick (1979) proposed the concept of enacted environment as a replacement for the external environment since 'the human actor does not react to an environment, he enacts it'. In which case, the category between organisation and environment is not clear and the environment is created and defined by individuals (Brooks and Weatherston, 1997). Thus, objective measures of the environment become irrelevant (Corrieria,

1996). In Porter's (1980a) and Ohmae's (1982) books, they use examples that some companies begin a price war against the profitability of the other companies within the industry. On the other hand, Weick's (1979) concept of enacted environment can be viewed as a development of the perceived environment since its intention is to understand the subjective approach to organisations' interactions with the environment.

Given the conceptual frameworks described above, the principal focus of the empirical research was on the perceived business environment and enacted business environment (see Figure 3.3 shown above). In this research context, the enacted business environment refers to strategies adopted by service companies operating in East Asia.

The decision to concentrate on the perceived and enacted business environment rests on two assumptions. First, research into the objective business environment for the oil and gas service organisations has received a huge amount of attention from industrial analysts or experts. Second, and equally importantly, few examples of work concerning the perceived and enacted business environment in the energy service sector in East Asia have been found.

Furthermore, with regard to the enacted business environment, the focus lies in the study of business strategies employed by energy service organisations operating in East Asia. There are three reasons for this and two are based upon Hofer's assumptions (1975).

First, the study of business strategy requires a smaller, less complex set of variables than the study of corporate strategy. Second, a firm cannot achieve long-term success at a corporate level until it knows how to achieve success at a business level (Hofer, 1975). Third, it was discovered that international energy service organisations had a vital role within the industrial sector in East Asia due to their advanced technological competencies, yet many of them had only business or operating units in the region. Hence, the study of strategy at a business level would be more appropriate and applicable.

3.2   Perceived business environment

The following will review the broader literature of previous studies in the field of the perceived business environment and business strategies and then explain the reasons for conducting relevant research.

3.2.1 Importance of analysing the business environment

The features of a firm's external environment are described in numerous ways. Most strategy theories characterise environments in terms of their levels, for example general, industrial, task or operating environment (Duncan, 1972; Pearce and Robinson, 1997) and firm-specific variables (Koberg, 1987; Daft et al., 1988; Ireland et al., 1987; Miller, 1993; Hegarty and Tihanyi, 1999; Luo and Park, 2001; Tan and Litschert, 1994). Thompson (1967) suggests that the priority for an organisation is to deal with the uncertain eventualities of the environment, particularly those of the task environment (Dill, 1958).

Early in the 1950s, organisational theorists started to investigate the organisation–environment interaction (Tung, 1979) and found that the views of managers play a central role in learning about the environment (Hegarty and Tihanyi, 1999). Strategic management theories and practices stress that understanding the business environment is crucial in every organisation's life. Managers are encouraged to become more responsive to the dictates of the external environment and are required to scan and assess environmental conditions when making strategic decisions (Fahey and Narayanan, 1986).

Fahey and Narayanan suggest that assessment implies identifying and evaluating how and why current and projected environmental changes affect or will affect strategic management of an organisation. Assessment attempts to investigate what key issues are presented by the environment and what the implications of these issues are for the organisation. Accurate assessment of the environment by managers may help bring about more effective strategies and thereby higher performance for long-term success (Downey et al., 1975; Hambrick, 1982; Daft et al., 1988; Hegarty and Tihanyi, 1999).

In spite of the fact that most publications on organisational or strategic management theories have introduced the concept of the environment, comprehensive analyses or empirical studies of environmental characteristics are limited.

Tung (1979) argues that a major obstruction has been how best to describe and conceptualise organisational environments. The reliability of an instrument for measuring the business environment is still to be developed and tested. In particular, research on managerial perceptions of the business environment remains an important theoretical and empirical task.

3.2.2 A typology of environmental dimensions

In the past, researchers have made a distinction between the composition of organisational environments and environmental characteristics or dimensions (Tung, 1979). The composition of environments refers to the factors or sectors encompassing economics, government regulations and policies, technology, society, customers, competitors and suppliers (Miles and Snow, 1978; Miller, 1993).

Environmental characteristics or dimensions refer to the aspects of the environment confronting the organisation, for example complexity, dynamism and hostility (Tung, 1979; Miles and Snow, 1978; Miller, 1993). The organisational environment could be static or dynamic, complex or simple, hostile or favourable (Mintzberg et al., 1998).

Perceived uncertainty

The concept of uncertainty has long been a central component of a number of theories of organisation and strategy. In recent years, researchers have devoted their attention to managerial perceptions under uncertain environmental conditions. The concept of perceived environmental uncertainty advanced by many researchers (Tung, 1979; Hrebiniak and Snow, 1980; Koberg, 1987, Daft et al., 1988) has been a key aspect of a number of strategy theories (Miles and Snow, 1978; Lawrence and Lorsch, 1967; Mintzberg et al., 1998; Miller, 1993).

Managers operating in the external environment context confront a variety of uncertain factors. International business management researchers have focused primarily on the assessment of government policy and macroeconomic uncertainties and appropriate organisational responses. Some researchers in the strategy field emphasise that uncertainties are related to market demand for products or services, product and process technologies, the availability of critical inputs, and strategic actions by competitors and potential entrants (Miller, 1993).

Most commonly, general environmental uncertainty includes the uncertainty of politics, economics, social or cultural factors and technology (PEST). Industrial or task or operating environmental uncertainty encompasses suppliers, buyers, potential entrants, substitute products or services and rivalry among competitors. Some analysts (Sutcliffe and Huber, 1998) have identified industry environment characteristics in terms of industry concentration, entry barriers and changes in demand or changes in product characteristics. The category of firm-specific uncertainties pertains to the uncertainties of operations, management, research and development as well as employee actions (Miller, 1993).

Research (Ireland et al., 1987; Kotha and Nair, 1995; Sutcliffe and Huber, 1998; Sutcliffe and Zaheer, 1998; Elenkov, 1997; Simerly and Li, 2000) integrating the perspectives on organisational uncertainties has been developed. As Miller (1993) finds out, 'a major obstacle to empirical research on perceived environmental uncertainties is the lack of well-established measurement instruments. Existing measures from organisation theory suffer from conceptual problems and inadequate reliability and validity.' That difficulty still exists.

Perceived complexity

Fahey and Narayanan (1986) define complexity as referring to the degree of similarity or differentiation between elements or entities within and across environmental factors or components. It pertains to the number and heterogeneity or diversity of factors and components in the external environment. Tung (1979) explains that the heterogeneity or diversity includes two arrays of variables: the number of factors and components in external environments, and the relative differentiation or variety of these factors and components. Early work (Lawrence and Lorsch, 1967; Duncan, 1972; Tung, 1979; Downey et al., 1975; Dess and Beard, 1984; Luo and Park, 2001) found that if the number and diversity of environmental factors or components increase, executives' cognitive abilities to figure out the significance are increasingly limited. As a result, the level of perceived environmental uncertainty increases.

Perceived dynamism

Previous research has directed attention to managerial perceptions under changing environmental conditions and suggested that dynamism is an important dimension (Lawrence and Lorsch, 1967; Duncan, 1972; Tung, 1979; Dess and Beard, 1984). Industry dynamics have been viewed as giving rise to managerial uncertainties. They suggest that environmental dynamism is the product of several forces operating at one time, including the growth of the size and number of organisations within an industry, and the growth of the rate of technological change and its dispersion throughout the industry.

Dess and Beard (1984) define environmental dynamism as the rate and degree of instability of environmental change. Rate of change refers to the frequency and enormity of the turbulence of environmental factors and components. Instability may increase the complexity of environmental factors and require prompt organisational action (Hegarty and Tihanyi, 1999).

Miles and Snow (1978) emphasise that uncertainty refers to the unpredictability of environmental or organisational variables that have an impact on corporate performance. An effect of increasing levels of environmental dynamism is to reduce access to knowledge needed to make strategic decisions. This, in turn, reduces the stability and predictability perceived by executives regarding environmental factors or components (Tung, 1979).

Tung (1979) defines that the concept of dynamism pertains to rate of change, which includes frequency and magnitude of change, and the stability of change or predictability of the change pattern. If the change is more or less random rather than following a trend, it may be too sudden and completely unpredictable for organisations to acquire the capability to deal with it. It was hypothesised that this sort of change would greatly increase the degree of environmental uncertainty perceived by executives (Tung, 1979). When it is difficult or impossible for an organisation to predict the latest changes and grasp the implications for operations and activities, the dynamism dimension thus has an impact on the degree of uncertainty perceived by executives (Thompson, 1967; Duncan, 1972; Downey et al., 1975, Tung, 1979; Simerly and Li, 2000). Simerly and Li (2000) propose that greater environmental uncertainty is associated with greater environmental dynamism.

Perceived hostility

Research has shown that there are two aspects to hostility (Luo and Park, 2001). Firstly, it points out how critical the resources controlled by each environmental sector are and, secondly, it refers to the deterrence factor, in other words, the extent to which each environmental sector becomes a threat to the growth of an organisation.

In the first case, hostility shows the extent to which resources required by the organisation are available in its environment and describes the capacity of the environment to support organisations in the marketplace (Fahey and Narayanan, 1986). In the second case, Pfeffer and Salancik's (1978) resource dependency theory proposes that organisations arrange their external relationships in response to the uncertainty arising from dependence on components of the environment. Using Tan and Litschert's (1994) approach and suggestions from the resource dependence perspective, hostility focuses on the degree of the organisation's dependence on others for resources. The degree of dependency on factors and components affects or restricts executives' management capability to carry out business activities.

In addition to the above debates, the level of perceived environmental hostility depends not only on resource availability, but also on relationships with direct environmental agencies and competition within the same industry (Mintzberg et al., 1998). As the environment becomes less favourable or more hostile, firms are subjected to greater uncertainty (Tan and Litschert, 1994; Luo and Park, 2001; Kotha and Nair, 1995).

In this study, some of these concerns have been addressed by examining relative effects of the business environment on the offshore oil and gas service sector in East Asia. The empirical research work studied executives' perceptions of the environmental characteristics in order to modify, on an empirical basis, an environmental typology grounded in organisational and strategic management theory.

A multidimensional construct (Elenkov, 1997; Luo and Park, 2001) has been developed to conceptualise environmental uncertainty. Detail on developing this instrument will be given in Section 3.6.

3.3   Concept of strategy

A most serious semantic problem in the strategic management literature is the lack of consistency in the use of fundamental concepts like strategy and business strategy. In some instances, strategy is defined as the means to achieve a goal; in others, the definition is expanded to include both the goal and the means to achieve it. In spite of the surrounding polemics, it can be seen that any differences in definitions lie mainly in their scope: some authors bring goals into the definition; others do not. This section reviews the overall traditional concepts of strategy and outlines the difficulties in reducing strategy concepts to one single definition.

3.3.1   A brief review of definitions

Different definitions of strategy have been addressed worldwide, from the East to the West, and from ancient times to the modern epoch.

Although strategy definitions vary, the major content of strategy concepts can be generalised. A generic definition provided in the *Concise Oxford Dictionary of Current English* (1995) says:

> Strategy is the art of war; is the management of army in a campaign, or the art of moving troops, ships, aircraft, etc. into favourable positions, or an instance of this or a plan formed according to it; is a plan of action or policy in business or politics etc.

The word 'strategy' in Middle English came from French *stratagème* and Latin *strategema* from Greek *stratēgēma* via *stratēgeō* to *stratēgo*. 'Stratagem is a cunning plan or scheme especially for deceiving an enemy or trickery' (*Oxford Dictionary*, 1995). From Greek literature, the term 'strategy' is a derivative of the word *stratēgo*, which means the psychological capability and skills that generals should have when they lead an army.

As a word, 'strategy' can be derived from two main sources: military and commercial terms. In military terms, the *Oxford English Dictionary* (1933) defines strategy as

> the art of a commander-in-chief; the art of projecting and direct-ing the larger military movements and operations of a campaign; usually distinguished from tactics, which is the art of handling forces in battle or in the immediate presence of the enemy.

The importance of strategy as a topic of study may be traced back to 1513 when Niccolò Machiavelli, the first great political phi-losopher of the Renaissance, studied how a government can stay in power. Later, Carl von Clausewitz (1780–1831), a famous Prussian military thinker and strategic theorist, provided a detailed concept of strategy. In his definition, he emphasises not only the content of strategy but also the process involved in it. In his book *Vom Kriege*, Clausewitz argues that

> strategy is the employment of the battle to gain the end of the war; it must therefore give an aim to the whole military action, which must be in accordance with the object of the war; in other words, strategy forms the plan of the war; and to this end it links together the series of acts which are to lead to the final decision ... it makes the plans for the separate campaigns and

regulates the combats to be fought in each ... strategy must go with the army to the field in order to arrange particulars on the spot, and to make the modifications in the general plan which incessantly become necessary in war. (*Vom Kriege* [Clausewitz], translated by Gatzke, 1942)

In commercial terms, Johnson and Scholes (2000) say that:

Strategy is the direction and scope of an organisation over the long term. It ideally matches its resources to its changing environment and in particular its markets, customers or clients so as to meet stakeholder expectations.

Strategy is about an organisation's ability to utilise its strengths and weaknesses to take advantage of the opportunities and overcome the threats facing the business. Strategy is about balancing the internal factors with the external drivers of the business.

Similarly, Kay (1999) suggests that 'business strategy is concerned with the match between a company's internal capabilities and its external environment' and says that strategy 'is a set of analytic techniques for understanding and influencing a company's position in the market place'. In his holistic review of definitions, Chandler (1962) describes strategy as the determination of the basic goals and objectives of an enterprise; and the adoption of courses of action and the allocation of resources necessary for carrying out these goals.

In short, strategy is the art of management in establishing goals, making decisions and preparing action plans with available capabilities and resources to gain the final victory in battles, and particularly, long-term success for a commercial purpose.

3.3.2   Major streams of the strategy concept

Military and political rules of strategy

Militarily, the first documented concept of strategy goes back to the world's first acknowledged strategist Sun Zi (also called Sun Tzu in the West), whose *The Art of War* was written in the later years of Spring and Autumn Period (approximately from 770 to 476 BC) in China and at almost the same time as the ancient Greek strategists.

The Art of War comprises 13 chapters. From Chapters 1 to 6, it is concerned with strategy. The first chapter, 'Laying the Plan', is the key to *The Art of War*. It begins, 'military action is of vital importance to the state. It is a matter of life and death, a road either to safety or to ruin.' Then Sun Zi points out that a victory depends not only on martial power but also other issues such as politics, economics, time, location, people and discipline. He emphasises that 'these elements should be familiar to every general; he who knows them will be victorious; he who does not know them will fail'. Both ancient Chinese and Greek strategists emphasise the attributes of guile and cunning required to develop a winning strategy.

Entry of strategy concepts into commerce

Through the medieval period into the present day revolutions in agriculture and industry have brought an increasing need to adapt strategy from a military use for commercial purposes (Alexander, 1990). Strategic warfare has been applied over many centuries to help kings win wars over their enemies. Low and Sirpal (1995) comment that the business world is like war. Military manoeuvres can similarly and equally be applied strategically to help businesses win the competition (Low, 2001; Low and Sirpal, 1995; Low and Lee, 1997).

Many business leaders and researchers have begun to bring the lessons of the battlefield into the marketplace. The relationship of military strategy and business elements is a major theme found in many strategy studies. Whenever possible, a preamble on the parallels between the military and business is always included in the study of corporate strategy (Alexander, 1990). For example, attempts were made by Low and Sirpal (1995) to extend the 36 Chinese classical strategies of war into the realms of Western generic business (Porter, 1980b, 1985; Hofer and Schendel, 1978) and corporate strategies (Drucker, 1974) for use in the business world.

3.3.3   Evolution of strategy: a chronological review

This section provides a compendium chart depicting the evolution of strategy. It moves chronologically to provide a snapshot view of the major influences on the concept of strategy. In order to do so, the period of strategy evolution is divided into four major parts (Figure 3.4).

Period I: initial appearance of strategy (post-Second World War era)

The first period refers to the years preceding the Second World War when strategy was used mostly as the military parlance for warfare

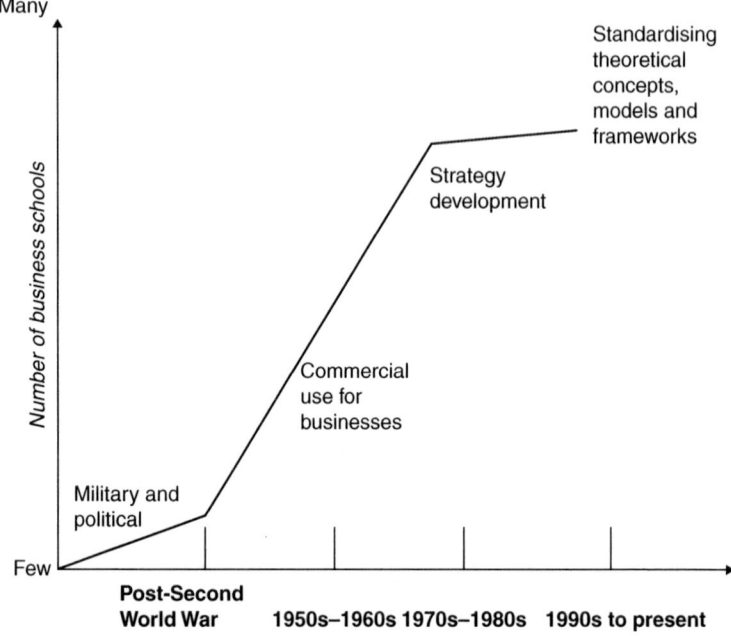

Figure 3.4   Evolution of strategy

and political control. As stated earlier, the recorded history of strategy reaches back to 476 BC when the Chinese military strategist Sun Zi wrote *The Art of War*, which has been regarded as the most influential classical strategic thinking in East Asia (Chen, 1995). Zhuge Liang (around AD 229) is another very famous Chinese military strategist, but the gist recommended by some researchers is that his book is in favour of generals displaying leadership, character, responsibility and knowledge. Therefore, Zhuge Liang's strategic thoughts are more suitable for managers who wish to have competent management skills.

The third highly influential ancient Chinese military strategy is 'Thirty-Six Chinese Classical Strategies of War'. It was written by an anonymous Chinese scholar (around AD 1644), gathering ancient Chinese strategists' thoughts emerging about 1500 years ago. On the whole, Low and Sirpal (1995) divide the entire book into six sections, each containing six strategies: strategies when in a superior position; strategies for confrontation; strategies for attack; strategies for confused situations; strategies for gaining ground; strategies for desperate situations. They suggest that the first three sections are useful when one holds the advantage and the second set useful when one is in a disadvantaged situation.

No matter what is stressed, the aim of strategy at this period was to assist army commanders to win the battles of the war and gain access to extended agricultural resources such as land for political rulers.

In the era of economic development between the First and Second World wars, there was an overlapping period when military strategies were transformed or applied for the use of business purposes. Over the period, for instance, the Japanese scholars studied and applied Sun Zi's strategic thinking to their management and business strategy-making. Low (2001) states that nowadays the ancient Chinese strategies of war are commonly applied to business practice by Asian managers.

However, although the first acknowledged strategy book appeared in China, strategy development has been generated mainly in the West.

Period II: theory building (1950s–1960s)

The second peak period of strategy development was during the 1960s. Many of the concepts that form the basis of today's understanding of strategy development were developed during the first half of the twentieth century. Since the concept of strategy was not developed extensively in the business literature until the late 1950s, almost no empirical research related to it was done until the early 1960s. Since the late 1960s, a number of studies have emphasised the generic content of an organisation's strategy.

At the early stage of the theory development, scholars like Ansoff (1965) laid the foundations for strategic planning by demonstrating the need to match business opportunities with organisational resources and illustrating the usefulness of strategic plans. Later, Hofer (1975) discovered that the concept of organisational strategy during this period had emerged as one of the cornerstones of both management theory and practice.

Formal theory in the 1960s focused on tools and techniques to help managers with decisions about business direction (Whittington, 1999). The critical works include the portfolio matrices produced by consultants like the Boston Consulting Group. By then industrial economics was dominated by the structure–conduct–performance paradigm. This emphasised how market structure was the principal influence on a company's behaviour (Kay, 1999). Feurer et al. (1995) summarise that this early phase was followed by a stage of generalisation in which researchers attempted to identify common patterns of success.

Period III: theory debate and development (1970s–1980s)

In the 1970s, both theory and research on the content of business and corporate strategy developed in piecemeal fashion. However, the most influential criticism made by Hofer (1975) is that much of this work has failed to differentiate between business and corporate strategies. Shortly after that, Miles and Snow (1978) developed a typology for assigning different business strategic orientations.

In the 1980s, the focus shifted from strategic planning towards strategic management. In 1980, Michael Porter's *Competitive Strategy: Techniques for Analysing Industries and Competitors* was published. Led by Porter (1980b, 1985), a broad range of concepts and techniques evolved which were aimed at building and sustaining competitive advantage by anticipating and exploiting business opportunities. This was popular during the 1980s and remains deeply entrenched in strategy thinking today. It translates the structure–conduct–performance paradigm into an industry structure model that can be used by business, although some people, such as Lynch (2000), have pointed out the serious limitations of this model. For example, the model does not allow for the phenomenon of cooperation within an industry.

As Feurer et al. (1995) summarised, during the 1970s and 1980s researchers increasingly recognised that strategy development cannot be regarded as a simple design mechanism but that different strategy processes may exist in different organisations. The idea that there may be a gap between the intended and achieved strategy had also been raised. Numerous studies culminated in a large number of strategy tools and frameworks that are still used for analysis purposes today.

Period IV: theory test, modification and standardisation (1990s to the present and future)

The fourth period is from the 1990s to the present and future. In the 1990s, the concept of strategy was still developing and the literature was growing rapidly in every direction, such as strategic groups, value chains, key success factors, and other ideas. The representative work was conducted by Mintzberg et al. (1998, 1999) by studying various types of strategies. Intended strategy and realised strategy mean that organisations develop plans for their future, in part by evolving patterns out of their past. Deliberate strategy and unrealised strategy refer to the intentions that are fully realised or not realised. Emergent strategies, where actions are taken one by one and some sort of consistency or pattern emerged over time, are not expressly intended.

Today, strategists become aware of the increased speed of change and the level of uncertainty in business environments. In fast-changing circumstances, some (Feurer et al., 1995) argue that it is not possible to determine a strategic direction for an organisation on a systematic basis. Hence, organisations must constantly adapt to such circumstances and move towards dynamic strategy development. As Alexander (1990) has pointed out, future strategy development will become more fragmented, more flexible, and enable local divisions to react decisively to the newest or strongest threat arising from the environment. However, there is a need to set up standards for universal application of numerous theoretical models, frameworks and theories.

3.4 Major streams of business strategy

Over the years, researchers identified more and more types of business strategy through both empirical and theoretical research, culminating in a wide range of business strategy models and typologies. Most of these business strategy models and typologies stand for two major schools of business strategy: Miles and Snow's taxonomy (1978) and Porter's theory (1980b, 1985).

3.4.1 Miles and Snow's typology of business strategic directions

Miles and Snow (1978) propose that organisations develop relatively enduring prototypes of strategic orientations in line with characteristics like the range of products and markets, technology solutions, desired growth pattern, and attitude toward change. They defined four major strategic trends for organisations.

According to Miles and Snow (1978), defenders seek stability and control in their operations to maximise efficiency. They provide a narrow range of products or services to serve well-defined niche markets. Normally, they focus on a single core technology and typically grow by market penetration. In contrast, prospectors provide a broad range of products or services and create changes in the industry. They pursue growth through product and market development. Analysers emphasise both stability and flexibility, and follow keenly the most promising changes developed by prospectors and defenders. Reactors are naturally unstable in terms of their adjustment and lack consistency in strategic options. As a result, the reactors' performance is relatively poor.

Researchers such as Hambrick (1982) criticise the Miles and Snow typology on the grounds that it is not the most elaborate framework that could be chosen. A later study by Wright et al. (1990) suggests a good-performing amalgamation strategy: the balancer. They assume that a balancer combines the features of defender, prospector and analyser.

Parnell et al. (2000) summarise a balancer as having three separate product-market spheres. The feature in the first sphere is very similar to that of defenders whose managers stress the domain of existing products or services and customers. The second sphere resembles the analyser type. Conducting technological change is encouraged in order to imitate the best of products and markets developed by prospectors. In the third sphere, managers attempt to initiate changes within the industry. The strategically operational style tends to bear a resemblance to the characteristic of prospector organisations.

The majority of research on generic business strategy has been conducted in relation to Western businesses. A limited number of studies have been conducted in East Asia. Little or nothing is known about the solutions needed for service companies within the oil and gas industry in China, Singapore and Malaysia. Empirical academic studies examining the competitive position for these service companies are also rare.

One of the research aims in this study attempts to narrow such a gap left by previous researchers. A conceptual framework of five types of business strategies, namely, defender, prospector, analyser, balancer

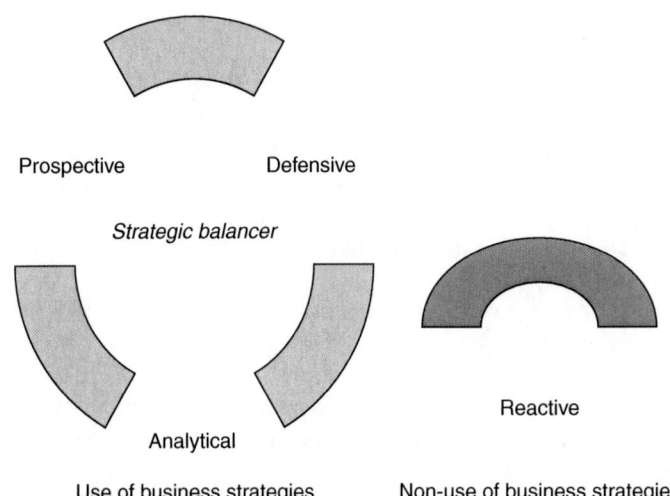

Prospective   Defensive

Strategic balancer

Analytical

Reactive

<u>Use of business strategies</u>   <u>Non-use of business strategies</u>

Figure 3.5 Modelling generic business strategies
Source: Based on Miles and Snow (1978).

and reactor, was developed and utilised for the empirical investigation (Figure 3.5). This conceptual framework will be presented in more detail later in Section 3.6.

3.4.2 Porter's taxonomy of generic competitive strategies

According to Porter (1980b), businesses can attain significant and enduring competitive advantage over their rivals by adroitly pursuing generic strategies (i.e. cost leadership, differentiation, low cost focus or differentiation focus and multiple strategies). A firm failing to adopt its strategy in any one of the generic directions or engaging in each generic strategy without achieving any of them will stay in the position of being 'stuck in the middle' (Porter, 1980b, 1985).

The first competitive strategy is overall cost leadership. Cost leadership is the total control of expenditure and pursuit of cost reductions for an efficient performance within an organisation so that competitive advantages such as a high relative market share or favourable access to raw materials can be achieved.

The second competitive strategy is differentiation. This generic strategy is concerned with bringing differences to the product or service produced by firms. It intends to create a unique feature that is perceived industry wide.

Focus is the third competitive strategy. This generic strategy is to gain competitive advantage through a company focusing on a particular customer group or industry-market segment or geographic market, and not attempting to serve any others. Competitive advantages can be gained if one's strategy is tailored to the needs of a relatively small niche market. A firm can achieve either differentiation or low cost or both.

Because the nature of this study accentuates business rather than marketing attributes, and in so far as no existing standards have been found to define the 'focus' level for firms within the oil and gas service sector, the focus strategy is not included for the empirical study. Consequently, attention has been devoted to the other three types of generic strategies.

The fourth competitive situation is 'stuck in the middle'. 'Stuck in the middle' is where a firm fails to develop its strategy in at least one of the three strategic directions stated above. The firm is regarded as one of low profitability. It is assumed that, in some industries, when some firms become stuck in the middle, it may mean that 'the smaller (focused or differentiated) firms and the largest (cost leadership) firms are the most profitable, and the medium-sized firms are the least profitable' (Porter, 1980b).

A number of empirical studies have been conducted to test the validity of Porter's (1980b, 1985) generic strategies. The debates lie in two principal areas. In the first area, empirical investigations contradict the prospect on the pursuit of more than one generic strategy (Miller, 1992; White, 1986). A key point is that low cost and differentiation strategies are incompatible. Many researchers support the argument that, for higher business performance, an organisation has to choose either low cost or differentiation as a prior strategy, rather than both (Porter, 1985; Faulkner and Bowman, 1995).

In the second area, some other researchers argue against Porter (1985) in his assertion of using a single generic strategy. They emphasise that a firm can pursue low cost and differentiation simultaneously and the combination of both strategies can also result in higher business performance (Chan and Wong, 1999; Proff, 2000).

Regardless of the above debates, a number of researchers (Helms et al., 1997; Yamin et al., 1999; Wright et al., 1990) support both schools of thought and suggest that organisations should deploy generic competitive strategies that best suit their circumstances.

3.4.3 Bowman's strategy clock: competitive positions

Porter (1980a, b, 1985) emphasises that the basis of generic business strategy is how customer or client needs can best be met, usually through achieving a certain competitive advantage. A competitive position is the basis on which a business might achieve advantages to outperform its competitors in a marketplace (Johnson and Scholes, 1999). Great attention has been paid to analysing generic strategies and competitive positions associated with organisational performance (Yamin et al., 1999).

According to Porter (1985), competitive advantage is the underlying concept of generic strategies. All organisations are in a competitive position in relation to each other as they compete either for customers or for resources (Johnson and Scholes, 1999). This is the case which can be applied in the oil and gas service industry: firms compete mainly for customers (i.e. operators) and increasingly more for the downstream-side supply resources such as available capacity, experienced personnel or reliable equipment.

In order to understand an organisation's strategic position within which it attains competitive advantage, Bowman's strategy clock (Johnson and Scholes, 1999), which is advanced by Faulkner and Bowman (1995) as a customer matrix, has been developed further in this study.

Two basic dimensions of the strategy clock are defined as perceived customer added value (PCAV) and perceived price. The PCAV is conceptualised by Faulkner and Bowman (1995) as perceived use value. In this study, the PCAV refers to senior managerial perceptions of the degree of the value that their businesses create to meet the satisfaction of customers. Perceived price refers to the level of price charged by firms for their products or services.

In order to simplify, five basic competitive positions (Figure 3.6) are defined as:

- Position 1 – high value, premium (i.e. above the moderate level) price;
- Position 2 – high value, moderate price;
- Position 3 – high value, competitive (i.e. below the moderate level) price;

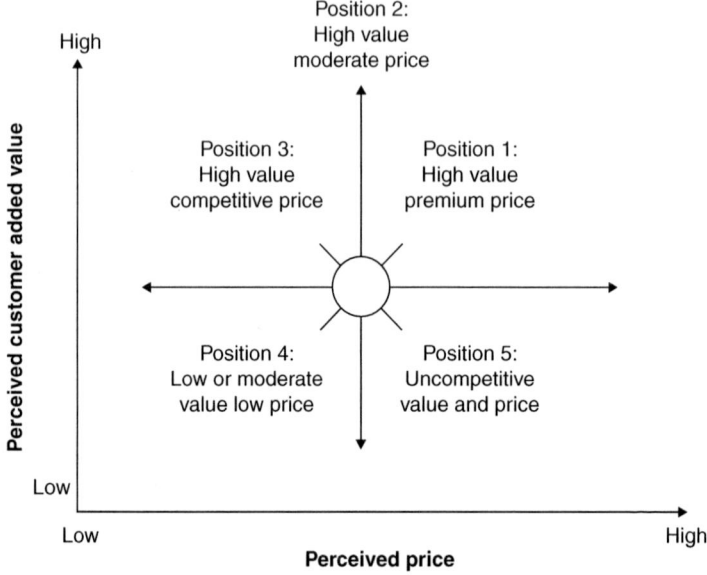

Figure 3.6   Strategic positions in a competitive market
Source: Based on Faulkner and Bowman (1995).

- Position 4 – low or moderate value, low price;
- Position 5 – uncompetitive value and price.

3.5   Conceptual development of strategic performance

Performance has been of interest to both academic researchers and for assessment of practising managers. There is a widely available literature for the organisational or business performance and there are numerous ways to measure performance (Nash, 1983).

Slater and Olson (2000) suggest that performance is a complex multidimensional construct that is influenced by both the level of analysis (i.e. individual, business unit, or organisation as a whole) and strategy type. However, Croteau and Bergeron (2001) argue that measuring organisational performance can be a problem because no universally recognised measure of this concept is available. Today, academic debates about issues of terminology, level of analysis, and conceptual bases for assessing performance are still ongoing.

The most dominant model in empirical research on business or organisational performance is based on the simple outcome of financial indicators and is referred to as the financial performance (Yamin et al., 1999). For oil and gas-related energy service companies, typical financial indicators usually include revenues or production turnover, backlog, profitability and total asset growth.

Some strategy studies have also employed market or value-based measurements in conjunction with financial based measures because of their relevance regardless of types of strategy (Parnell et al., 2000). In addition to indicators of financial performance, another broader conceptualisation of business performance (Venkatraman and Ramanujam, 1986) would include non-financial emphasis indicators such as operational performance (Helms et al., 1997). The available measures can be product or service quality, effectiveness, efficiency, added value and technological reliability.

In order to measure the fulfilment of the overall goals of a business, using just the single functional performance approach was considered an unbalanced solution applied in this study. The concept has therefore been expanded to a wider range, including the indicators which are assumed to reflect the general achievement of an organisational business. This concept is referred to as strategic performance (Figure 3.7), which covers overall management activities such as organisational image, innovation, marketing, operations, finance, and human resource management.

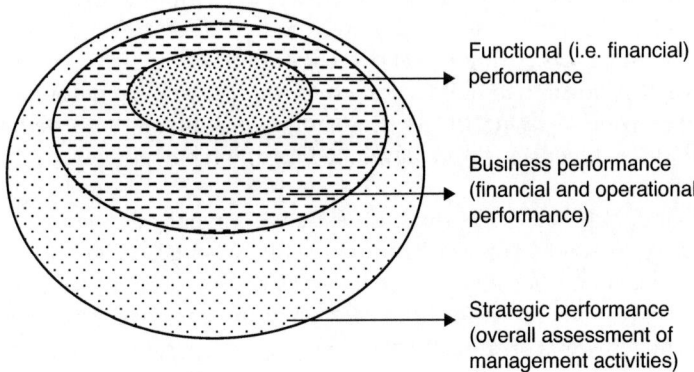

Functional (i.e. financial) performance

Business performance (financial and operational performance)

Strategic performance (overall assessment of management activities)

Figure 3.7   Strategic performance measurement model

3.6   Developing theoretical frameworks

3.6.1   The assessment of environmental sectors

Previous research (Tan and Litschert, 1994; Miles and Snow, 1978; Hrebiniak and Snow, 1980) has developed a number of sectors or categories to measure the three dimensions of environmental uncertainty in the general and task environment. However, the limitation of the validity of their environmental components is that the scope of the concept on each of the environmental factors is still too broad for Chinese managers to understand perfectly its meaning. For instance, when assessing the regulatory environmental sector (Tan and Litschert, 1994), Chinese managers might wonder which levels of governmental regulations they should consider. Is it the international, national or local level?

Several studies (Hegarty and Tihanyi, 1999; Luo and Park, 2001) have been conducted to model the perceived business environment through quantitative analysis, but few consider how to combine the degree and scope of determinant environmental sectors and environmental dimensions to gain a comprehensive perception of the business environment. As such, it is indeed necessary to develop a well-defined typology for interpreting and analysing the three environmental dimensions (i.e. complexity, dynamism and hostility) under scrutiny.

Based on the work of Simerly and Li (2000), Sutcliffe and Zaheer (1998) and Elenkov (1997), the measurement of the three environmental dimensions focused on the six environmental sectors, namely, economics, regulatory, technology, customers, suppliers and competitors. Using the above environmental components as a basis, it is possible to build a scale that categorises the environment and provides a rating for its degree of environmental uncertainty. The three dimensions were developed for the typology presented in Table 3.1.

A model for assessing the business environment has been generated in this study. Three environmental dimensions, viz. complexity, dynamism and hostility, were defined as the measures for assessing the environmental sectors at a task environment level. The senior executives' views of, and attitudes toward, the environmental factors were measured by using a 45-item scale.

Table 3.1 The business environment assessment model

Dimensions	Variables	Homogeneity		7-point bipolar scale				Heterogeneity
	Environmental sectors							
(I) Complexity	Economics	Homogeneity	1 2 3 4 5 6 7					Heterogeneity
	Technology		1 2 3 4 5 6 7					
	Regulatory		1 2 3 4 5 6 7					
	Customers		1 2 3 4 5 6 7					
	Competitors		1 2 3 4 5 6 7					
	Suppliers		1 2 3 4 5 6 7					
(II) Dynamism	Economics	Predictable	1 2 3 4 5 6 7					Unpredictable
	Technology		1 2 3 4 5 6 7					
	Regulatory		1 2 3 4 5 6 7					
	Customers		1 2 3 4 5 6 7					
	Competitors		1 2 3 4 5 6 7					
	Suppliers		1 2 3 4 5 6 7					
(III) Hostility	Economics	Benign	1 2 3 4 5 6 7					Hostile
	Technology		1 2 3 4 5 6 7					
	Regulatory		1 2 3 4 5 6 7					
	Customers		1 2 3 4 5 6 7					
	Competitors		1 2 3 4 5 6 7					
	Suppliers		1 2 3 4 5 6 7					

The uncertainty level: Very low   Very high

The research survey questions (see Appendix) were arranged as multi-item scales corresponding to the six environmental sectors. Each of the environmental items indicates the degree of environmental complexity, hostility and dynamism, and in turn, the level of environmental uncertainty. Hence, four basic assumptions emerged as follows:

- The greater the numbers or heterogeneity (diversity) of an environmental sector, the higher the degree of the perceived complexity of this environmental sector.
- The higher the unpredictability of an environmental sector, the higher the degree of the perceived dynamism.
- The higher the degree of the difficulties of resource availability and resource deterrence, the higher the degree of the perceived hostility.
- When the degree of the perceived complexity, dynamism and hostility is high, the perceived environmental uncertainty also tends to be high.

Managerial perceptions on the perceived complexity and hostility were measured by a series of 7-point bipolar (the 'bipolar' concept was advanced by Osgood in 1957) rating scales (Zikmund, 2000). Bipolar adjectives anchor the right and left of the scale, with 1 indicating that the respondent most strongly agrees with the left assessment, 7 indicating that the respondent most strongly agrees with the right assessment, and 4 showing that the respondent felt that, for his organisation, the situation was midway between both. Similarly, environmental dynamism was examined by the perceived predictability of these six environmental sectors. The answers were measured by using a 7-point numerical scale, with 1 indicating very predictable, 7 very unpredictable and 4 a neutral situation.

For instance, to measure the attitude towards perceived dynamism, from left to right, the scale intervals were interpreted as very static, static, tend to be static, neutral, tend to be dynamic, dynamic and very dynamic. As such, a question may ask:

The business environment in which you operate is

very static   **1   2   3   4   5   6   7**   *very dynamic*

Weights of 7, 6, 5, 4, 3, 2 and 1 were assigned to the answers and a score was assigned to each position on the rating scale. Strong agreement indicates the most favourable attitudes toward the statement, and the weight of 1 was assigned to the most positive response while 7 was assigned to the most negative response. Hence, in each measuring situation, a score has been assigned for the perceived business environmental dynamism. Should '5' be selected by a respondent, it indicates that the respondent agrees with the assessment as stated on the right.

3.6.2 Category of strategies by service companies

Figure 3.8 illustrates a balancer framework for an energy service organisation. A defender pursues a concentrated growth across the entire oil and gas industry, ranging from serving exploration and appraisal, to development and production, and to decommissioning; and across different regional markets home and abroad.

Prospectors pursue market development through finding new markets for their existing products and services. They are also actively engaged in product/service development through a focus on differentiation and therefore new products/services can be available for existing clients.

An analyser may have concentric or conglomerate business units. In addition to a mixed feature of combining defender and prospector,

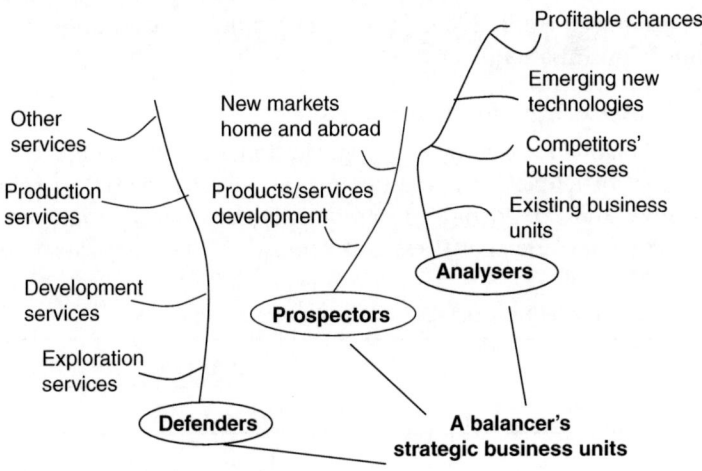

Figure 3.8   Balancer in the oil and gas service sector

it also benchmarks the changes of identified competitors' businesses and seeks to obtain potential profitable or new technologies from oil companies, technological institutions or government parties. A balancer is a combination of the above three strategic types. It may contain businesses outside the oil and gas industry.

3.6.3   Taxonomy of competitive strategies

In this study, a taxonomic framework for generic business strategies has been generated based on conceptual theories and contemporary empirical studies. Four basic generic strategies are defined as low-cost, differentiation, hybrid and no-purpose (Table 3.2).

As Yamin et al. (1999) demonstrate, differentiation strategy aims at achieving, even at considerable cost, a superior quality throughout the value chain and creating the image of a unique feature; while the emphasis of a low-cost strategy is on lowering cost more than competitors wherever possible.

A hybrid strategy means that a firm seeks to deploy more than one of the generic strategies and achieves cost leadership and differentiation simultaneously (Johnson and Scholes, 1999; Proff, 2000; Porter, 1980a, b).

If a firm fails to develop its strategy in at least one of the three directions, or is inconsistent in pursuing the generic strategies and achieves no competitive advantage, the firm has no distinctive strategy (Campbell-Hunt, 2000). Such firms are described as no-purpose strategy organisations.

3.6.4   Strategic performance assessment

It is difficult to access directly organisational archive data such as financial or marketing figures as they are usually viewed as being sensitive and confidential. Consequently, this study employed a subjective measurement (Dess and Robinson, 1984) that calls upon

Table 3.2   Competitive strategies matrix for service companies

Lowering cost	*Differentiating business*	
	Yes	*No*
Yes	Hybrid	Low-cost
No	Differentiation	No-purpose

Source: Based on Porter's (1980a, b) taxonomy.

managerial perceptions. The reliability of this self-reporting approach has been proved by various studies (Tan and Litschert, 1994; Luo and Tan, 1998; Luo and Park, 2001).

However, a 5-point Likert scale (from bottom 20 per cent to top 20 per cent in the industry) developed by Tan and Litschert (1994) was inappropriate because the relevant information for the oil and gas service sector in China, Singapore and Malaysia was unavailable or hard for respondents to obtain. To solve this, a 7-point interval scale was drawn from Ramanujam and Venkatraman's (1987) work, with 1 indicating much less or worse and 7 indicating much more or better (see Appendix, Survey Questionnaire).

3.6.5   The ESP model

The research for establishing the link between environmental uncertainty, strategy and performance has long been asserted in conceptual work in strategic management (Miles and Snow, 1978; Porter, 1980a, b). More recently, empirical evidence of the existence and the nature of this link in various management disciplines has been explored. In their work, Kotha and Nair (1995) examine the roles played by the environment and realised strategies on performance. Similar research conducted by Tan and Litschert (1994), Parnell et al. (2000) and Ramanujam and Venkatraman (1987) have studied the relationship between strategic planning and performance. Venkatraman and Presott (1990) also test the proposition of a positive performance impact of environment–strategy coalignment.

A broader literature search is now shifting towards the relationships between the business environment, strategy and performance. Most work has been carried out to examine strategy and performance correlations. Researchers (e.g. Wright et al., 1990; Croteau and Bergeron, 2001; Parnell et al., 2000) have not only attempted to classify business strategies into typologies but also studied more effectively relations between strategy and other variables such as performance.

A common observation is that the more specific the type of business strategy adopted by an organisation, the better the organisational performance. However, there should be a negative link for the reactor type, which means that balancers are high performers while reactors are low performers (Parnell et al., 2000).

In their work, Croteau and Bergeron (2001) also discover that there is a positive link between strategic activities and performance for prospectors, whereas there is a negative link for reactors. Slater and

Olson (2000) focus on the evidence that prospectors, analysers and defenders can achieve superior performance when implementing strategies appropriately.

In this field of studies, empirical evidence has also been gathered in China. Tan and Litschert (1994) find that defensive-oriented strategies are related to higher overall performance. Luo and Tan (1998) further prove that both defender and analyser strategies are positively and significantly related to the Chinese firms' financial performance. Luo and Park (2001) provide evidence showing that analyser-oriented organisations produce a high performance, while prospector and defender orientations lead to poor financial performance when mismatching with a highly dynamic and complex Chinese market.

Moreover, Porter (1985) suggests that a firm achieving cost leadership or differentiation may potentially obtain above-average profitability performance. The rewards of simultaneously achieving cost leadership (which implies lower costs) and differentiation (which leads to premium prices) are also great because the firm gains extra benefits (Porter, 1980a, b).

Several researchers provide evidence to support Porter that a company adopting its strategy in a hybrid direction outperforms businesses pursuing a single generic strategy (Chan and Wong, 1999; Proff, 2000; Wright et al., 1990). Researchers (Feurer et al., 1995) have realised that business performance depends not only on the formulation and successful implementation of a given strategy but also on the process by which competitive positions are created or maintained. In their opinion, successful performance can be achieved through forming the basis of differentiation.

Previous work has discovered that the business environment in an East Asian country like China is dynamic (Tan and Lischert, 1994) and competition within the oil and service industry in the region is fierce. Hambrick (1982) points out that, in a dynamic industry environment, the low-cost strategy would be unlikely to be found. Moreover, for a firm pursuing the lowest cost, information on cost levels of competitors is usually very difficult to obtain (Faulkner and Bowman, 1995).

The strategic situation of a firm is extremely poor when it develops its strategy in a way that has no distinctive emphasis (Porter, 1980b, 1985). Porter further suggests that such a firm is almost guaranteed low profitability because firms deploying low-cost, differentiation or hybrid strategies are able to sustain a stronger position and compete better in any segment.

However, researchers (Porter, 1980b; Faulkner and Bowman, 1995) also point out that, although a firm may have no distinctive emphasis on any generic strategies for achieving competitive advantage, it can still attain satisfactory performance, notably through earning attractive profits. There may be two reasons for this. First, the structure of its industry is highly favourable or secondly, the other firms in the industry do not successfully adopt any generic strategies. Nevertheless, in the long term, those firms that have not made a choice between alternatives of generic strategies will be exposed when an industry becomes mature and competition is fierce (Faulkner and Bowman, 1995).

Based on these debates, three assumptions can be relevant for further investigations:

- For oil and gas service companies in East Asia, a low-cost generic strategy may not be a preferred option for senior management.
- Having a generic strategy such as low-cost or differentiation or hybrid yields the firm a higher level of strategic performance; and pursuing none of these three generic strategies (no-purpose) produces a relatively poor strategic performance.
- Hybrid organisations outperform those competing mainly with either low-cost or differentiation strategy.

Furthermore, Helms et al. (1997) have pointed out that a firm's success is associated with the possession of strategic advantages, rather than strict adherence to Porter's generic strategies. They insist that the more competitive advantages a firm holds the better the business performance it achieves. Hence, one more assumption can be explored further:

- Organisations that aim for the position of high value with competitive price outperform companies in other categories of competitive positions.

Finally, several hundred empirical studies in strategic management have examined the correlations between strategy and performance but most work has been found to investigate strategy–performance by excluding the role played by the business environment (Kotha and Nair, 1995). Li's research (2000) on Chinese township enterprises provides evidence that their businesses have been forced to develop coping strategies to deal with the significant level of uncertainty and adversity of the business environment. Few researchers (Venkatraman and Prescott, 1990; Tan and Litschert, 1994; Luo and Park, 2001) have attempted to provide empirical evidence of the environment–strategy–performance (ESP) paradigm.

As empirical studies examining the relative impact of strategy and environment on performance in East Asian industries are rare, this study explores the interface between environment and strategy, and the coalignment between environment and performance. It also examines the typology of business strategy and correlates this with the assessment of managerial perceptions of strategic performance.

3.7   Summary

A general objective guiding this study was to examine the common trends concerning strategic orientations across the existing oil and gas service organisations in the East Asian business environment. In this study, the term 'environment' refers to the external environment that comprises all forces outside an organisation.

Within this conceptual context, the scope of the present research is limited to three focused aspects: the business environment assessment, the identification of existing strategies and evaluation of strategic performance.

The principal focus of the empirical research was on the perceived and enacted business environments. The perceived environment is concerned with managerial perceptions on the external business environment in which they operate in China, Singapore and Malaysia. The enacted business environment refers to strategies adopted by the service companies operating in the three selected countries.

In this study, a multidimensional construct was employed to conceptualise environmental uncertainty. Three dimensions of the environment, viz. complexity, dynamism and hostility, were discussed in the assessment of environmental sectors. A conceptual framework of five types of business strategies, namely, defender, prospector, analyser, balancer and reactor, was developed for the empirical investigation. An organisational performance assessment model was generated under a concept called 'strategic performance' covering management activities at all levels within an organisation. A theoretical framework on the coalignment and variation between environment and performance, environment and strategy, and strategy and performance was also developed for result analysis.

4
Perceived Business Environment in China, Singapore and Malaysia

This chapter demonstrates empirical findings on the business environment for the oil and gas service sector in the three East Asian countries studied. It contains the background context of the oil and gas service companies involved in this study and managerial perceptions on the business environment in which they operate. The correlations among environmental dimensions and the differences or similarities of the managerial perceptions on the business environment across the three countries are presented. In this chapter, nonparametric techniques are applied mainly for analysis.

4.1 Background context

Questions related to the organisational background are based on a category scale. The type of data measured is nominal and therefore relevant types of descriptive analysis are frequency tables and proportions. The results are presented in three areas: company profiles, industrial segments of the energy service sector and business activities in East Asia.

4.1.1 Company profiles

This part of the questionnaire includes category data which are measured by a nominal-level scale. Frequency SPSS (statistical package for the social sciences) subprograms are used to present the raw count of cases for each value.

Table 4.1　Participating organisations summary (N = 98)

	Frequency	%
Legal status category		
Division/subsidiary companies	41	41.8
Independent companies	39	39.8
Operating/business units	18	18.4
Country where based		
Singapore	39	39.8
China	31	31.6
Malaysia	21	21.4
Other countries	7	7.1
Full-time employees		
1–49	39	39.8
50–199	20	20.4
200–499	19	19.4
500–2999	16	16.3
3000 or more	4	4.1
Years since present business was formed		
5 Years or more	92	93.9
<5 Years	6	6.1
Ownership category		
Wholly foreign owned	38	38.8
Joint venture	22	22.4
Wholly domestic private/individual	15	15.3
Domestic shareholding/public limited companies	13	13.2
Wholly domestic state owned	10	10.2
Respondent's position		
Senior manager	62	63.3
Regional manager or business head	30	30.6
Functional head	6	6.1

Of the 98 participating organisations (Table 4.1), 39 were independent companies, 41 were divisional or subsidiary companies, 18 were operating or business units and of these, one was a regional office, with 31 organisations situated in China, 39 in Singapore, 21 in Malaysia and 7 in other countries such as Thailand, Indonesia and the United Kingdom. Those companies located outside China, Singapore and Malaysia operated businesses in these three countries, but their business strategic decisions were made at the headquarters elsewhere.

Regarding the organisational size in terms of full-time employees, 39 organisations had fewer than 50 full-time (or equivalent) employees while 16 were large or very large size organisations with 500 or more than 500 employees. Most of the participating organisations were classified as medium and large organisations with 50 or more employees (61.2 per cent).

In China, a small enterprise refers to one which is owned individually, or by a collective of working people, who participate in the democratic management of their own workplace. In many countries, organisations with fewer than 500 employees are defined as small and medium-sized enterprises; and organisations with fewer than 50–100 employees are defined as small enterprises (Lu, 1999). In this study, the small size enterprises refer to organisations with fewer than 50 employees; medium-sized enterprises refer to organisations with employees from 50 to 499; and large size organisations refer to those with 500 or more employees.

A large majority of the organisations had operated for five or more years since their existing businesses were formed (93.9 per cent). Only six participating organisations had operated businesses for less than five years.

Most of the respondent organisations were wholly foreign owned or joint ventures (61.2 per cent); a considerable proportion were domestic organisations (38.8 per cent), including wholly domestic state owned, wholly domestic private or individual owned, and domestic shareholding or public limited companies. Three of the domestic organisations indicated that their ownership is 'PTY Limited', proprietary, and foreign majority with domestic private. PTY Limited is the normal designation for private limited liability companies in Singapore.

The ownership category result shows the fact that these service organisations were either international or global. The majority within the oil and gas service sector in East Asia were linked to foreign investment.

Regarding managerial positions of the participating respondents, senior managers (i.e. managing directors, general managers, chief executive officers or presidents) accounted for 63.3 per cent. Just above 30 per cent were managers at an organisational middle management level (i.e. regional sales or operating or division managers, organisational business development managers). In the context of this study, these middle-level executives were regional heads and can be regarded as at a senior management level in the region of

East Asia. Besides, a small proportion, 6.1 per cent of responses, were from functional heads such as organisational marketing or sales or operations or commercial managers.

At this stage, it was decided to use the 98 participating organisations for environmental analysis. However, when doing strategy-related data analysis, the six functional managers' responses would be excluded; and when doing strategic performance-related analysis, 11 organisations that had less than five years in their existing businesses and had functional managers' responses would be excluded.

4.1.2   Industrial segments of the energy service sector

Adopting the concept generated by Simmons and Company International (1999), a model for assigning service companies to different oil and gas service segments was developed. From the simple SPSS bar chart, the results pertaining to the participating organisations' service activities within the energy industry are reported in Figure 4.1.

Of the 98 participating organisations, most focused their businesses on serving the upstream oil and gas industry (67 per cent); nearly half provided services to other oil and gas service companies (44 per cent); 21 per cent were involved in serving midstream oil and gas transportation; and 33 per cent in serving downstream oil and gas refining or processing and marketing. The business involvement with respect to energy services outside the oil and gas industry is considerable. Of these service organisations, 7 per cent served the hydropower industry, 13 per cent served the electrical industry, 5 per cent served nuclear and 8 per cent served other energy sectors such as wave, coal and wind power.

In addition, for the 98 participating organisations' upstream service activities (Figure 4.2), most of them served oil production (67 per cent), 43 per cent served exploration and 39 per cent provided services to development (including completions). The smallest proportion was engaged in serving appraisal activities (14 per cent). Noticeably, a minority of the participating organisations served other oil and gas sectors outwith upstream activities (18 per cent).

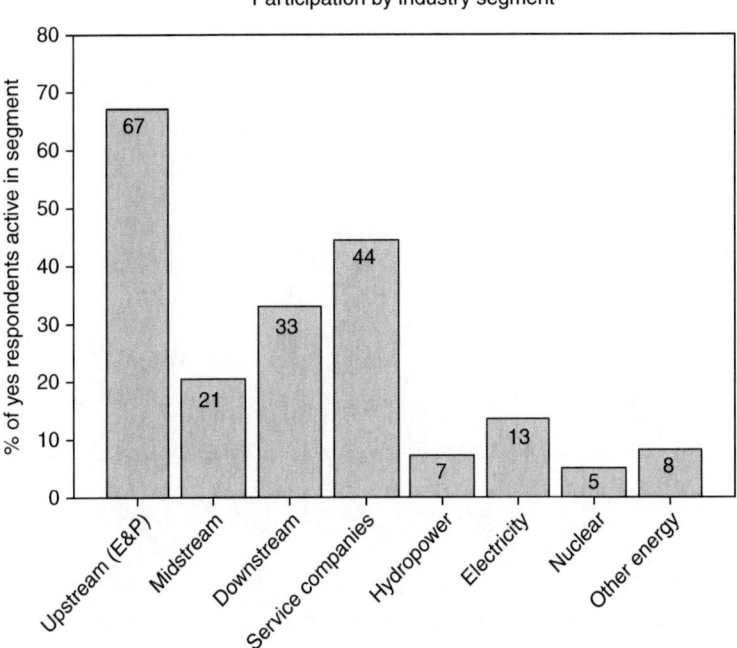

Figure 4.1 Service segments in the energy industry

4.1.3 Geographical presence in Asia

The service activities of the participating organisations were widely distributed in various countries and regions throughout East Asia (Figure 4.3).

For one organisation, its businesses may be carried out in more than one country. Consequently, the majority of the 98 participating organisations had operated businesses mainly in China, Singapore, Malaysia and Indonesia (64, 53, 60 and 54 per cent respectively). A considerable proportion of the organisations had conducted their services in the Philippines, Vietnam and Thailand (31, 38 and 42 per cent respectively). A small proportion of them carried out businesses in Japan and South Korea (13 and 15 per cent respectively). In addition, 14 per cent of these organisations had operated businesses

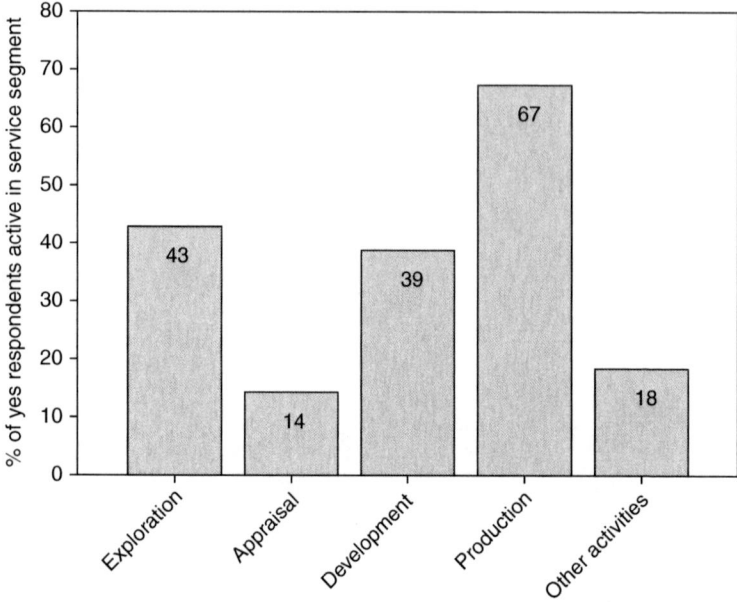

Figure 4.2   Oil and gas upstream service activities

in other East Asian countries or regions such as Brunei, Taiwan, Hong Kong, Maldives, Myanmar and Azerbaijan.

4.1.4   Years in business

From Figure 4.4, most of the organisations ten or more years old were from Singapore; whereas for the organisations between five and ten years in age, most were from China.

A common trend shows that, in each of the three countries, a big majority of the oil and gas service organisations were more than five years old, with 97 per cent in both China and Singapore and 90.5 per cent in Malaysia.

The proportion of the Singapore-based organisations which were more than ten years old is bigger (74 per cent) than the proportion of the China and Malaysia-based organisations at the same age (58 and 67 per cent respectively). Since the proportion of organisations with less than five years in their existing businesses is small,

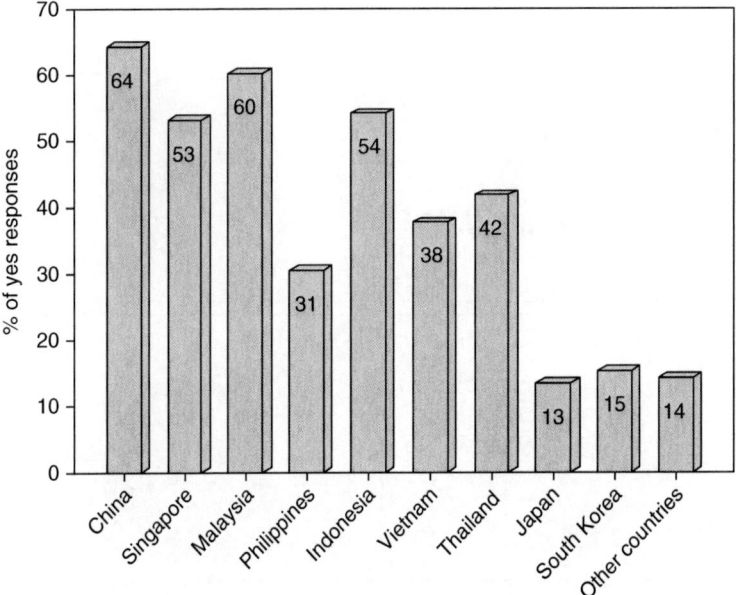

Yes responses of business in East Asia

Figure 4.3 Geographical presence of survey companies

it is considered that the statistical results on managerial perceptions pertaining to strategic performance will be reliable if this category is excluded.

In general, the offshore oil and gas service industry of Singapore appears to have a longer history than that of China and Malaysia. Meanwhile, for organisations which were between five and ten years old, the proportion in China is bigger (39 per cent) than the proportion in Singapore (23 per cent) and Malaysia (23.8 per cent). This supports the fact that offshore oil and gas service businesses were being carried on earlier in Singapore than in China.

4.1.5 Ownership

In Table 4.2, most of the 21 joint ventures come from China (61.9 per cent) and noticeably, all ten domestic state-owned companies are from China. A big majority of the 38 wholly foreign-owned

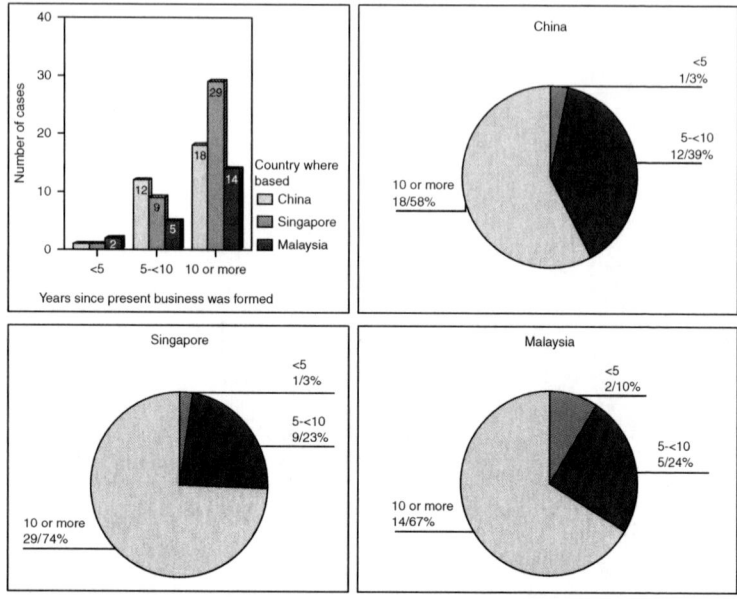

Figure 4.4   A comparison of organisational age

entities are from Singapore (75.8 per cent) and 6 out of 13 domestic shareholding organisations or public limited companies are from Malaysia. Half of 14 domestic private or individual enterprises are from Singapore.

Looking at the situation in individual countries, joint ventures in China's oil and gas service sector are more popular (41.9 per cent) than those in Malaysia (23.8 per cent) and Singapore (7.7 per cent). This shows that a joint venture (JV) is a preferred approach which can be adopted in China. Hence, an effective entry strategy for foreign service organisations to enter into China's market is to set up a JV with Chinese partners.

Wholly foreign-owned entities are dominant within the offshore oil and gas service industrial sector in Singapore (64.1 per cent). For Malaysia, the types of ownership for oil and gas service organisations are fragmented: the service sector is a mixture of wholly domestic private companies (19.0 per cent), wholly foreign-owned

Table 4.2 Ownership category – country cross-tabulation

			Country where based			Total
			China	Singapore	Malaysia	
Ownership category	Wholly domestic state owned	Count	10			10
		% within ownership category	100.0%			100.0%
		% within country where based	32.3%			11.0%
		% of Total	11.0%			11.0%
	Wholly domestic private/individual	Count	3	7	4	14
		% within ownership category	21.4%	50.0%	28.6%	100.0%
		% within country where based	9.7%	17.9%	19.0%	15.4%
		% of Total	3.3%	7.7%	4.4%	15.4%
	Wholly foreign owned	Count	2	25	6	33
		% within ownership category	6.1%	75.8%	18.2%	100.0%
		% within country where based	6.5%	64.1%	28.6%	36.3%
		% of Total	2.2%	27.5%	6.6%	36.3%
	Joint venture	Count	13	3	5	21
		% within ownership category	61.9%	14.3%	23.8%	100.0%
		% within country where based	41.9%	7.7%	23.8%	23.1%
		% of Total	14.3%	3.3%	5.5%	23.1%
	Domestic share holding/public limited company	Count	3	4	6	13
		% within ownership category	23.1%	30.8%	46.2%	100.0%
		% within country where based	9.7%	10.3%	28.6%	14.3%
		% of Total	3.3%	4.4%	6.6%	14.3%
Total		Count	31	39	21	91
		% within ownership category	34.1%	42.9%	23.1%	100.0%
		% within country where based	100.0%	100.0%	100.0%	100.0%
		% of Total	34.1%	42.9%	23.1%	100.0%

entities (28.6 per cent), JVs (23.8 per cent) and domestic shareholding organisations or public limited companies (28.6 per cent).

In contrast with Singapore and Malaysia, the Chinese government has played a business role in the offshore oil and gas service sector, as China is the only country where service companies were wholly state owned (32.3 per cent of the total participating companies in China); Singapore and Malaysia appeared to have a free competitive market with less governmental interference in oil service businesses. The results also indicate that Singapore is indeed a favourable place where foreign companies consider establishing their own regional headquarters (64.1 per cent of the total participating companies in Singapore).

4.2   Assessing task environment segments

The six environmental factors – political or regulatory, economic, technological, customers, suppliers and competitors – form a task environment confronting the daily and direct business opera- tions of oil and gas organisations. According to Proposition 1 demonstrated in Chapter 1, these factors will be perceived by oil and gas service executives to be significant for the growth of their businesses in East Asia. In order to investigate this proposition, the significance of an environmental factor is evaluated in two dimensions. The first dimension is its importance and the second its impact on the growth of the service organisation's businesses in East Asia.

In order to evaluate differences among the managerial percep- tions on the importance or impact of the six task environmental factors, the non-parametric Friedman tests were conducted. The tests were significant, ^{2} = 159.9 (df = 5), $p < 0.001$ for environmen- tal importance and χ^2 = 156.3 (df = 5), $p < 0.001$ for environmental impact.

Furthermore, the value of the mean rank for the importance of the customers sector is 4.97, the highest of the six environmental factors; the mean rank of the importance of suppliers is 2.22, the lowest of the six environmental factors.

For the mean ranks of environmental impact, the highest value is the customers' (4.82) and the lowest mean rank value is the

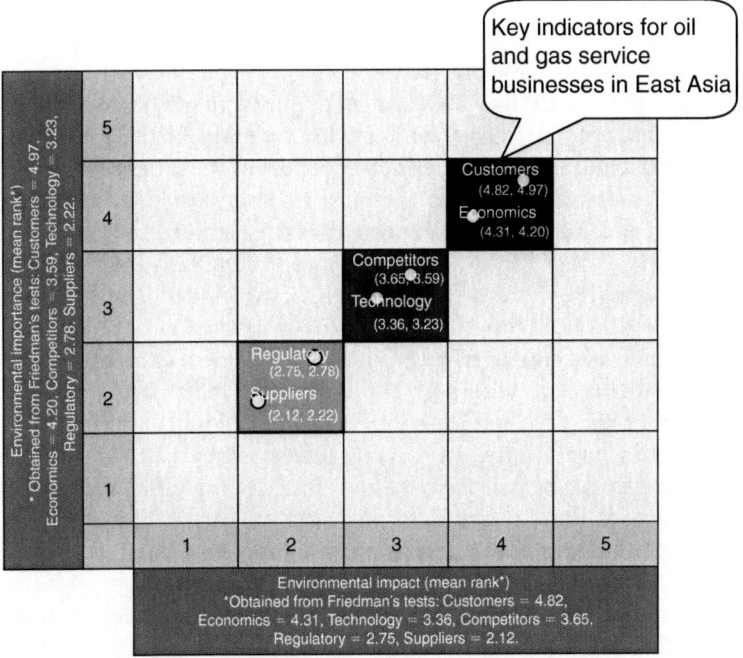

Figure 4.5 Key indicators driving oil and gas service business

suppliers' (2.12). Using these mean rank values of the environmental importance and impact, a hypothetical approach of a leading industrial indicators matrix can be generated in Figure 4.5.

It is observed that the rank of environmental impact is the same as the rank of environmental importance. Of the six environmental factors, the impact of the customers' factor was perceived to be the strongest, and this environmental factor was also perceived to be the most important. On the other hand, the impact and importance of the suppliers' factor were perceived to be the least.

Obviously, the managerial perceptions on environmental influences differed significantly. The extent of environmental impacts on the growth of their business in East Asia was found to vary. The impact of economic, customers' and competitors' factors was strong, and especially the customers' factor having the strongest impact. The technological, regulatory and suppliers' factors were perceived to be moderate and the suppliers' factor had the least significant impact on the businesses of these organisations.

With respect to the importance of environmental conditions, senior managers perceived some environmental sectors as having a higher level of importance than others. Again, the customers' factor was not surprisingly seen as the most important, while the suppliers' influence was the least important of the six task environmental factors.

Similarly, the results provide significant evidence to support the view that the economic, technological, political or regulatory, customers' and competitors' environmental factors are important. The results also reveal that the suppliers' sector is considered important for most service companies operating in East Asia, but its level of importance is relatively lower compared to other task environmental factors. Hence, the results support the assumption that each of the six task environmental factors is important for the growth of oil and gas service businesses in China, Singapore and Malaysia.

The initial significant results of the impact of the economic, technological, customers' and competitors' factors are selected by executives to be in the category of 'strong or very strong'. The result on the regulatory factor is, however, perceived as a moderate impact and for the suppliers' factor, executives perceived its impact on business to be in the 'non-existent or very weak and weak' category.

The observed results support the original assumption that customers strongly affect the growth of oil and gas service businesses in East Asia. For an assessment of suppliers' impact, the observed results tend not to support the original assumption that they have a strong impact on oil and gas service businesses. This might be the reason why the earlier results show that quite a number of executives regarded the suppliers' factor as slightly important as described above. The results also disprove the assumption that the regulatory factor would have a strong effect on the service businesses in East Asia. Rather, the observed results indicate that the impact of the regulatory factor is moderate. Hence, the results tend to support the assumption that the impact of the customers', economic, technological and competitors' factors would be strong for oil and gas service businesses in East Asia.

Consequently, senior management may devote most attention to customers when scanning and assessing the task environmental conditions for strategic decisions. From the emerged ranking positions presented in Figure 4.5 for each of the six task environmental sectors, the customers', economic, competitors' and technological factors can be regarded as the leading indicators within the oil and gas service sector in the three selected countries. In particular, oil clients-related data can be used to determine the required approaches of service delivery or the establishment of competitive positions for various business units. Such data include oil operators' preference for price, and their demand for existing and new products or services, or the demand for the quality of products or services.

Overall, Proposition 1 that six task environmental factors are significant for the growth of oil and gas service businesses in East Asia is supported by the research results.

4.3 Managerial perceptions on the business environment

This section attempts to investigate Proposition 2 that, for oil and gas service companies that operate in East Asian countries like China, Singapore and Malaysia, the nature of the business environment is uncertain. It also seeks to prove the contention of Proposition 3 that the business environment in which the service organisations operate in East Asia is dynamic, complex and hostile.

4.3.1 Preliminary observations of environmental dimensions

The SPSS output (Table 4.3) shows the table of descriptive statistics for the four variables, viz. complexity, hostility, dynamism and uncertainty, providing an insight into the perceptions of these environmental dimensions. As the analysis explores the nature of the perceived business environment, the mean is employed as a hypothetical value (Field, 2000) for the measure of central tendency.

From Table 4.3, on average, the participating executives perceived that the business environment in which they operated in East Asia was uncertain ($M = 4.17$), complicated ($M = 5.04$) and dynamic ($M = 5.22$). However, for the perception of environmental hostility, the environment tended to be benign ($M = 3.81$) for service organisations to conduct their businesses in the region.

Table 4.3   Perceptions of environmental dimensions in East Asia (N = 98)

	Perceived complexity	Perceived hostility	Perceived dynamism	Perceived uncertainty
Mean	5.04	3.81	5.20	4.14
Standard error of mean	0.13	0.15	0.13	0.15
Median	5.00	4.00	5.00	4.50
Mode	5.00	4.00	5.00	5.00
Standard deviation	1.26	1.50	1.28	1.53
Variance	1.59	2.26	1.63	2.33
Skewness	−0.58	−0.05	−0.88	−0.35
Standard error of skewness	0.24	0.24	0.24	0.24
Range	6.00	6.00	6.00	6.00
Minimum	1.00	1.00	1.00	1.00
Maximum	7.00	7.00	7.00	7.00

At this stage, the intention (demonstrated in Proposition 2) that the nature of the business environment is perceived to be uncertain is preliminarily supported. However, Proposition 3 has gained limited support from the preliminary results.

In the following paragraphs, factors associated with the perceived complexity, dynamism and hostility are examined.

For environmental uncertainty, the z-score of skewness is $-0.35/0.24 = -1.46$. For perceived dynamism, the z-score of skewness is $-0.88/0.24 = -3.67$. For environmental complexity and hostility, the z-scores of skewness are -2.42 and -0.21 respectively. As a z-value above 1.96 is considered significantly different from chance to be problematic (Field, 2000), it is clear that the dynamism and complexity scores are negatively skewed, indicating a pile-up of scores on the right of the distributions and hence most respondents gave high scores, except for hostility. Figure 4.6 presents bar charts of each of the four variables with the normal distribution overlaid.

The graphs displayed in Figure 4.6 represent various scenarios. Firstly, it looks as if the uncertainty score is somewhat normally distributed. In this sense, a few (4) respondents perceived the business environment as very uncertain and a few (6) perceived it as very certain, but half of the respondents' (49) perception fell into the uncertainty category.

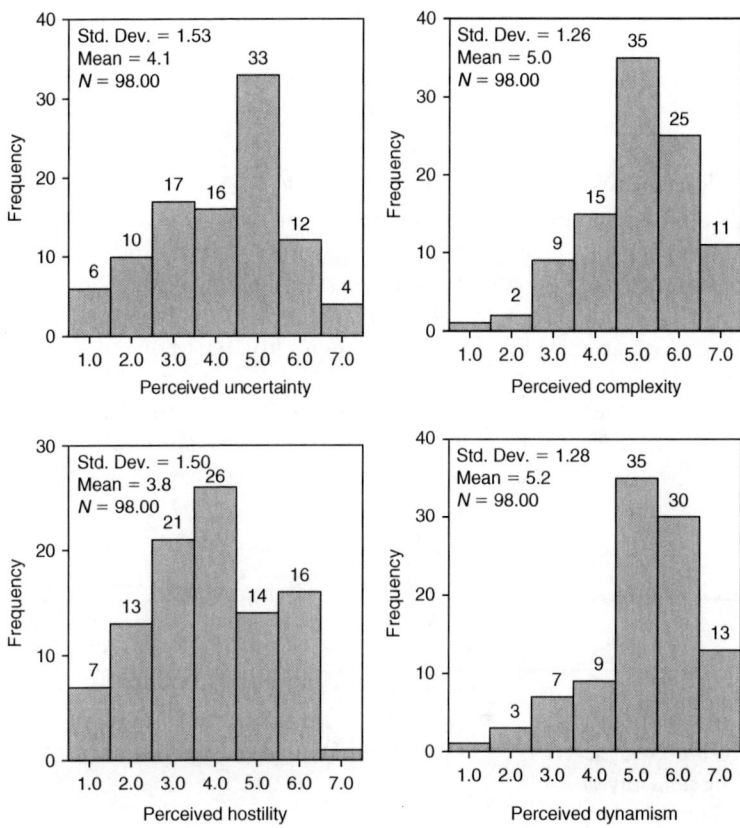

Figure 4.6 Nature of the business environment in East Asia

Secondly, the hostility scores are distinctive because this distribution is fairly clearly not normal and there are two peaks indicative of two modes. This suggests that the respondents' opinion about environmental hostility was divided: most perceived that the environment was pleasant (41.8 per cent); 26.5 per cent of the respondents felt it was in a neutral position between pleasant and unpleasant, whereas 21.6 per cent perceived that it was an unpleasant environment.

Finally, complexity and dynamism tests produced very negatively skewed data, indicating that the majority of the respondents perceived the business environment as complex and dynamic whereas the minority perceived it as simple and static. This observation corresponds with the earlier discussions about the z-scores.

4.3.2   Factors reflecting perceived complexity

A ten-item scale was used to assess the perceived environmental complexity. The minimum, maximum, mean and standard deviation were tabulated for each environmental sector. The standard deviation values presented in Table 4.4 are from 1.23, the lowest (i.e. economics knowledge required), to 1.80, the highest (i.e. number of suppliers).

These values are relatively high compared to the mean. The ratings for the assessment of environmental complexity are clearly spread from the mean, that is, for some respondents, very high ratings

Table 4.4   Perceived environmental complexity

	Minimum	Maximum	Mean	Standard deviation
Economics knowledge required	1.00	7.00	4.77	1.23
Technological level	2.00	7.00	5.09	1.36
The similarity of industrial products/ services	1.00	7.00	3.51	1.59
Government regulations, legislation and policies	1.00	7.00	4.34	1.51
Number of oil and gas clients	1.00	7.00	4.30	1.64
Needs and preferences of oil and gas clients	1.00	7.00	3.22	1.53
Number of suppliers	1.00	7.00	3.92	1.80
Similarity of supply conditions by suppliers	1.00	7.00	3.80	1.55
Number of firms within the industry sector	1.00	7.00	3.26	1.75
Scope of firms within the industry sector	1.00	7.00	5.03	1.59

(e.g. 7 indicating the greatest complexity of an environmental factor) were given and for others, the ratings were very low (e.g. 1 indicating the simplest situation of the same environmental factor). As the standard deviations are far from zero, the mean may be a poor fit of the data. In spite of this, the mean can be used as a hypothetical value.

An approach to interpreting the mean value is developed as follows. The weight of 1 is assigned to the response of the most favourable attitude on the statement anchored at the beginning (positive adjectives); the weight of 7 is assigned to the response of strongest agreement indicating the most favourable attitude on the statement anchored at the end (negative adjectives); 4 is the point assigned to the neutral attitude on that statement. If the mean is above 4, hypothetically, the perception is associated with the right statement showing the level of complexity; if below 4, it is associated with the left statement showing the extent of simplification. By doing this, the observed results emerge as follows.

Five environmental influences show the levels of complexity (Table 4.4):

- The knowledge required for understanding the economic situation in the region where they operated was complicated ($M = 4.77$).
- The government regulations, legislation and policies tended to be sophisticated ($M = 4.34$).
- The level of technology involved in the oil and gas service sector was high ($M = 5.09$).
- The number of customers whom they served within the oil and gas industry tended to be large ($M = 4.30$).
- The scope of companies within the service sector in which they operated was extensive as they came from all over the world ($M = 5.03$).

On the other hand, five other issues appear to show the simplification of the environmental influences. The observed results show that:

- The needs and preferences of the oil and gas clients whom they serve were similar ($M = 3.51$).
- Both the number of firms within the industry sector and the number of suppliers tended to be small ($M = 3.26$ and 3.92); in these circumstances, the niche market was shared or dominated mainly by a few firms.

- Supply conditions (e.g. price, quality, speed or service) provided by their suppliers tended to be similar ($M = 3.80$).

In addition, using a simple bar chart (Figure 4.7), more than half of the respondents thought that, within the service industrial sector in which they operated, the products or services offered to clients were similar to each other.

4.3.3   Factors reflecting perceived hostility

Thirteen items were used to assess the perceived environmental hostility, which was also measured by tabulating the minimum, maximum, mean and standard deviation. For the same reason as stated in the previous section, the appropriated measure of central tendency is the mean value. The method employed in that section was used to demonstrate the hypothetical means. The details on the distributions for each of these items could also be reported by using a simple bar chart. To simplify, the results are summarised in Table 4.5.

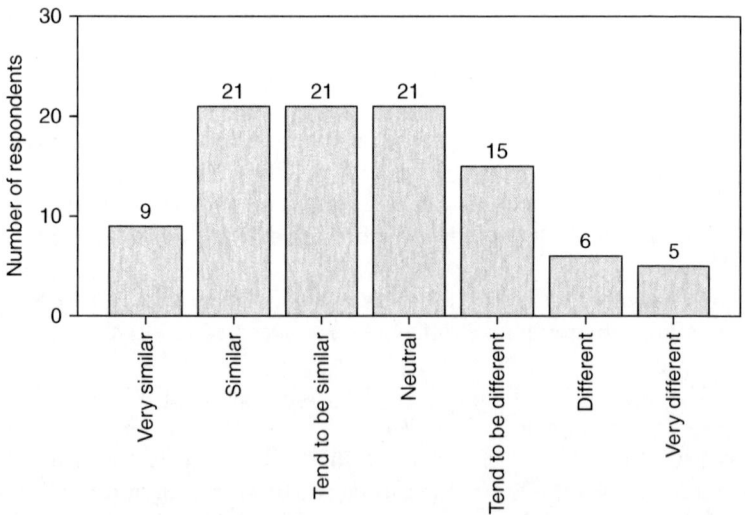

Figure 4.7   Similarity of industrial products or services

Table 4.5 Perceived environmental hostility

	Minimum	Maximum	Mean	Standard deviation
Size of market demand	1.00	7.00	3.32	1.48
Changing of market demand	1.00	6.00	3.34	1.19
Access to available technologies	1.00	7.00	3.73	1.44
National government regulations and legislation	1.00	7.00	4.03	1.36
Local government policies	1.00	7.00	4.14	1.31
Relationship with government	1.00	7.00	4.13	1.40
Levels of key customers switching to competitors	1.00	7.00	4.66	1.46
Relationship with customers	1.00	6.00	2.29	1.06
Access to suppliers for materials or goods	1.00	7.00	3.01	1.32
Relationship with suppliers	1.00	5.00	2.70	1.02
Entry barriers	1.00	6.00	3.14	1.37
Rivalry among competitors	1.00	7.00	4.54	1.49
Competitive actions	1.00	7.00	4.29	1.44
Relationship with competitors	1.00	7.00	4.17	1.44

Overall, the favourable situation was relevant to seven environmental influences:

- Market demand within the oil and gas industry which they served was big ($M = 3.32$) and also increasing ($M = 3.34$).
- It was easy to gain access to available technologies and to suppliers to obtain available raw materials or standard goods and services ($M = 3.73$ and 3.01).
- The participating organisations had established good relationships with their key oil and gas clients ($M = 2.29$).
- The relationships between the participating organisations and their key suppliers also tended to be supportive ($M = 2.70$).
- Entry barriers to the oil and gas service sector in which they operate were slightly high ($M = 3.14$), and this situation could discourage new competitors from entering into existing industrial sectors.

The following three environmental influences made the environment unfavourable to the oil and gas service organisations:

- It was easy for key customers to switch to another competitor's products or services ($M = 4.67$).
- Competitive actions adopted by some firms within the service sector might tend to be unreasonable ($M = 4.29$).
- Rivalry among the competitors within the service sector in which they operate was turbulent ($M = 4.54$).

As service firms within a niche industrial sector provide similar products or services to their common oil clients, competition was largely based on price. However, lowering price was not considered an appropriate way of obtaining contracts as it could cause quality problems. Operators in China preferred to take both price and previous quality performance into consideration when they evaluated bidding proposals.

Several environmental influences could be interpreted as having a neutral impact on service organisations. National government regulations and legislation in the region could benefit some organisations while limiting other firms' service businesses ($M = 4.03$); for those firms holding a neutral opinion, some governmental policies brought benefits whereas some restricted businesses.

Local-level government policies (e.g. Customs or administrative bureaux) could have both negative and positive influences on the oil service companies' businesses ($M = 4.14$). Some of the participating organisations had a close relationship with government, some were distant from the government, and others were neither distant nor close to the government ($M = 4.13$).

The relationship of some participating organisations with their competitors was collaborative, and the relationship of some other participating organisations with their competitors was uncollaborative, while some organisations had a neutral relationship (i.e. neither bad nor good) with competitors ($M = 4.17$).

A conspicuous observation emerges that the majority of oil and gas service organisations in East Asia had amicable relationships with their competitors, although the competition could be very strong.

4.3.4 Factors reflecting perceived dynamism

Eighteen items were used to measure the level of environmental dynamism and the means were used to provide hypothetical connotations of the data. From descriptive statistics shown in Table 4.6, apart from the fact that the influence of changes in competitive price appeared to be at a neutral point between predictable and unpredictable ($M = 4.04$), most of the factors in the six environmental sectors stayed about the same from year to year.

Table 4.6 Perceived environmental dynamism

	Minimum	Maximum	Mean	Standard deviation
Customer demand for existing products/services	1.00	7.00	3.01	1.39
Customer demand for new products/services	1.00	6.00	3.50	1.38
Customer demand for higher quality or more service	1.00	7.00	2.93	1.47
Customer preference for lower prices	1.00	7.00	2.31	1.39
Changes in competitive price	1.00	7.00	4.04	1.54
Competitors' quality improvement	1.00	7.00	3.51	1.18
Competitors' introduction of new products/services	1.00	7.00	3.69	1.39
Suppliers' rising prices	1.00	7.00	3.43	1.36
Suppliers' quality reduction	1.00	7.00	3.85	1.27
Suppliers' introduction of new materials or standard products	1.00	7.00	3.69	1.18
Changes in oil and gas E&P level	1.00	7.00	3.76	1.74
Changes in well counts	1.00	7.00	3.47	1.44
Changes in rig counts	1.00	7.00	3.64	1.49
Technological changes	1.00	6.00	3.26	1.11
Rate of technological diffusion	1.00	6.00	3.52	1.28
Changes in national regulations and legislation	1.00	7.00	3.76	1.51
Changes in local government policies	1.00	7.00	3.80	1.55

The predictable customer influences were:

- demand for existing products or services ($M = 3.01$) and for new products or services ($M = 3.50$);
- demand for higher quality or more services ($M = 2.93$);
- and customers' preference for lower prices ($M = 2.31$).

For the competitive sector, improvement in quality of products or services by competitors and competitors' introduction of new products or services appeared to be predictable ($M = 3.51$ and 3.69). Service organisations were able to predict their suppliers' rising prices ($M = 3.43$) and the reduction in quality of suppliers' goods and services ($M = 3.85$), as well as suppliers' introduction of new materials or standard products ($M = 3.69$).

For regulatory influences, the participating organisations appeared to have advance warning of changes in government regulations, policies and legislation at both national and local levels ($M = 3.76$ and 3.80).

The industry economic factors such as oil and gas exploration and production, well counts and rig activities in the East Asian region were predictable ($M = 3.27$, 3.64 and 3.27).

The service firms were also able to forecast technological changes in the service sector and the rate of technological diffusion throughout the sector ($M = 3.27$ and 3.52).

Obviously, the above results show that the unpredictability does not reflect the perceived dynamism. Senior executives perceived the environment as dynamic yet they were able to predict dynamic changes occurring in the six environmental sectors.

4.4   Associations of business environment dimensions

This part of the analysis was conducted from three major aspects. Firstly, it looks at whether the perceived environmental uncertainty is associated (positively or negatively) with the perceived environmental complexity, dynamism and hostility (Proposition 4). Secondly, it intends to examine how the perceived environmental uncertainty is associated with the influence of the variables within the six task environmental sectors (Proposition 5). Thirdly, it investigates how the perceived complexity, dynamism and hostility are associated with the various environmental variables (Proposition 6).

4.4.1 Correlations of uncertainty, complexity, dynamism and hostility

After a preliminary glance at the data, the correlation analysis was carried out. In order to examine further whether the perceived environmental uncertainty is associated with the perceived environmental complexity, hostility and dynamism, Spearman's rho (r) values were computed. The prediction is that the perceived uncertainty would correlate with the perceived complexity, dynamism or hostility. This hypothesis is not directional, so a two-tailed test (Field, 2000) was selected.

The results (Table 4.7) show that the perceived uncertainty has a 0.209 correlation with the perceived dynamism, and a 0.269 correlation with the perceived hostility. The correlation coefficients in these samples are low but significant ($p = 0.039$ and 0.007). The correlation between the perceived uncertainty and complexity is not significant ($p = 0.157$). Hence, when the perceived hostility and dynamism are high, the perceived uncertainty also tends to be high.

Another observation that can be found is that the perceived dynamism is associated with the perceived complexity ($r = 0.236$, $p = 0.019$) and the perceived hostility ($r = 0.234$, $p = 0.02$). In this sense, the higher the degree of perceived complexity or hostility, the higher the degree of perceived dynamism.

It has proved the assumption that the higher the degree of the perceived environmental dynamism and hostility, the higher the degree of perceived uncertainty. However, the results do not support the assumption that there is a relationship between the perceived complexity and the perceived uncertainty.

In the context of this study, not enough evidence can be given to prove that only the three environmental dimensions (perceived complexity, dynamism and hostility) can reflect the perceived uncertainty. Thus, the research results give limited support to Proposition 4. As a consequence, the perceived environmental uncertainty is measured further by assessing factors in the task environmental sectors below.

Table 4.7 Correlations of uncertainty, complexity, dynamism and hostility

Spearman's rho		Perceived complexity	Perceived dynamism	Perceived hostility	Perceived uncertainty
Perceived complexity	Correlation coefficient	1.000	0.236*	0.064	−0.144
	Significance (2-tailed)	–	0.019	0.530	0.157
	N	98	98	98	98
Perceived dynamism	Correlation coefficient	0.236*	1.000	0.234*	0.209*
	Significance (2-tailed)	0.019	–	0.020	0.039
	N	98	98	98	98
Perceived hostility	Correlation coefficient	0.064	0.234*	1.000	0.269**
	Significance (2-tailed)	0.530	0.020	–	0.007
	N	98	98	98	98
Perceived uncertainty	Correlation coefficient	−0.144	0.209*	0.269**	1.000
	Significance (2-tailed)	0.157	0.039	0.007	–
	N	98	98	98	98

* Correlation is significant at the 0.05 level (2-tailed).
** Correlation is significant at the 0.01 level (2-tailed).

4.4.2 Perceived environmental uncertainty and environmental factors

Spearman's correlation coefficients were computed between the perceived environmental uncertainty, numbers and heterogeneity (i.e. the complexity characteristics), unpredictability (i.e. the dynamism characteristics), and resource difficulties and deterrence (i.e. the hostility characteristics) of the six task environmental sectors.

The results of the correlation analyses presented in Table 4.8 show that 19 correlations were statistically significant at the 0.05 or 0.01 levels. The significant relationships between the environmental uncertainty and each of the individual environmental factors gave coefficients ranging from 0.221 to 0.401, showing low positive correlations.

With respect to the numbers and heterogeneity (or the emerged complexity) of the environmental sectors, the environmental uncertainty is associated with government regulations, legislation and policies ($r = 0.221$, $p = 0.029$) and the supply conditions by suppliers ($r = 0.226$, $p = 0.026$).

The results suggest that the more complicated the government regulations, legislations and policies in the region where a service organisation operates in East Asia, or the more different the supply conditions (e.g. price, quality, speed or service) provided by the organisation's suppliers, the higher the degree of the perceived environmental uncertainty.

Pertaining to the difficulties of resource availability and resource deterrence (i.e. the emerged hostility), the perceived uncertainty is associated with seven factors:

- key customers switching to competitors ($r = 0.244$, $p = 0.015$);
- customers' relationship ($r = 0.242$, $p = 0.016$);
- access to suppliers for materials or goods ($r = 0.245$, $p = 0.016$);
- relationship with suppliers ($r = 0.262$, $p = 0.01$);
- rivalry among competitors ($r = 0.276$, $p = 0.006$);
- competitive actions ($r = 0.209$, $p = 0.039$);
- competitors' relationships with each other ($r = 0.229$, $p = 0.023$).

Hence, the difficulties of resource availability and resource deterrence in the six task environmental sectors may be the occasion of

Table 4.8 Perceived uncertainty and environmental factors

Spearman's rho		Uncertainty
Uncertainty	Correlation coefficient	1.000
	Significance (2-tailed)	–
Government regulations,	Correlation coefficient	0.221*
legislation and policies	Significance (2-tailed)	0.029
Similarity of supply conditions	Correlation coefficient	0.226*
by suppliers	Significance (2-tailed)	0.026
Levels of key customers	Correlation coefficient	0.244*
switching to competitors	Significance (2-tailed)	0.015
Relationships with customers	Correlation coefficient	0.242*
	Significance (2-tailed)	0.016
Access to suppliers for materials	Correlation coefficient	0.245*
or goods	Significance (2-tailed)	0.016
Relationships with suppliers	Correlation coefficient	0.262*
	Significance (2-tailed)	0.010
Rivalry among competitors	Correlation coefficient	0.276*
	Significance (2-tailed)	0.006
Competitive actions	Correlation coefficient	0.209*
	Significance (2-tailed)	0.039
Relationships with competitors	Correlation coefficient	0.229*
	Significance (2-tailed)	0.023
Customer demand for existing	Correlation coefficient	0.307*
products/services	Significance (2-tailed)	0.002
Customer demand for higher	Correlation coefficient	0.270
quality or more service	Significance (2-tailed)	0.007
Changes in competitive price	Correlation coefficient	0.291*
	Significance (2-tailed)	0.004
Suppliers' quality reduction	Correlation coefficient	0.401**
	Significance (2-tailed)	0.000
Changes in oil and gas	Correlation coefficient	0.380**
E&P level	Significance (2-tailed)	0.000
Changes in well counts	Correlation coefficient	0.230*
	Significance (2-tailed)	0.024
Changes in rig counts	Correlation coefficient	0.283*
	Significance (2-tailed)	0.006
Rate of technological diffusion	Correlation coefficient	0.282*
	Significance (2-tailed)	0.005
Changes in national	Correlation coefficient	0.301**
regulations and legislation	Significance (2-tailed)	0.003
Changes in local	Correlation coefficient	0.250*
government policies	Significance (2-tailed)	0.013

*Correlation is significant at the 0.05 level (2-tailed).
**Correlation is significant at the 0.01 level (2-tailed).

business environmental uncertainty. If the level of perceived hostility (difficulty) of each of these seven environmental factors becomes higher, the degree of business environmental uncertainty perceived by the service executives also tends to be higher.

For environmental unpredictability, the perceived uncertainty is associated with ten environmental factors:

- customer demand for existing products or services ($r = 0.307$, $p = 0.002$);
- customer demand for higher quality or more service ($r = 0.27$, $p = 0.007$);
- changes in the competitors' competitive price ($r = 0.291$, $p = 0.004$);
- the reduction of suppliers' quality ($r = 0.401$, $p < 0.001$);
- changes in the oil and gas E&P level ($r = 0.38$, $p < 0.000$);
- changes in well counts ($r = 0.23$, $p = 0.024$);
- changes in rig counts ($r = 0.283$, $p = 0.006$);
- rate of technological diffusion ($r = 0.282$, $p = 0.005$);
- changes in national regulations and legislation ($r = 0.301$, $p = 0.003$);
- and changes in local government policies ($r = 0.25$, $p = 0.013$).

The above results suggest that any unpredictability of the changes emerging in the six task environment sectors (i.e. regulatory, technological, economic, customers, competitors and suppliers) may result in the uncertainty of the business environment in which service organisations operate. If the level of the unpredictability of each of these environmental factors becomes higher, the degree of business environmental uncertainty perceived by service executives also tends to be higher.

To conclude, four comments can be made. Firstly, the more easily the customers switch to another competitor's products or services, the higher the degree of environmental uncertainty.

Secondly, the perceived environmental uncertainty is also associated with the nature of organisational relationships. The better the relationships with clients, or the more supportive the relationships with suppliers, or the more collaborative the relationships with competitors, the higher the level of the environmental certainty perceived by the service executives.

Thirdly, when access to suppliers to obtain available raw materials or standard goods and services is more difficult, the perceived uncertainty also becomes higher.

Fourthly, when rivalry among competitors within the service sector in which an organisation operates is more turbulent, or competitive actions adopted by firms within the service sector are more unreasonable, managerial perceptions on the business environment tend to be more uncertain.

The above results provide evidence to support the assumption that the perceived uncertainty is associated with individual environmental factors. In general, the greater the diversity (in numbers of customers, suppliers and technology, etc.) associated with an environmental factor, the higher the degree of the perceived environmental uncertainty. The higher the degree of perceived environmental unpredictability, the higher the degree of the perceived environmental uncertainty. The higher the degree of the difficulties with regard to resource availability and resource deterrence, the higher the degree of perceived environmental uncertainty. Proposition 5 is therefore supported by these results.

4.4.3   Environmental dimensions and factors

Spearman's rho values indicating the perceived environmental complexity and associated environmental factors are presented in Table 4.9. The perceived complexity has a significant correlation with the variable of 'economic knowledge required' ($r = 0.396$, $p < 0.001$), a significant correlation with the variable of 'technological level' ($r = 0.386$, $p < 0.001$), and a significant correlation with the variable of 'customers' needs and preferences' ($r = 0.226$, $p < 0.001$). This is consistent with the preliminary findings obtained in Section 4.3.

There is a link between the perceived environmental complexity and the sophisticated economic knowledge required, the technological level and diversification of customers' needs and preferences. This means that the more sophisticated the knowledge required to understand the economic situation, the more complex the perceived business environment; the higher the degree of the level of technology involved in the oil and gas service sector in which an organisation operates, the more complex the perceived business environment; and the more different the needs and preferences of the oil and gas clients whom the organisation serves, the more complex the perceived business environment.

In conclusion, for the oil and gas service organisations in East Asian countries like China, Singapore and Malaysia, the perceived environmental complexity is associated with the level of technologies existing in the industry, the customer-related factors and oil economic conditions.

Spearman's rho values indicating the perceived environmental hostility and associated environmental factors are presented in Table 4.10. The results show that the perceived hostility has a positive correlation with the variable of 'relationships with customers' ($r = 0.354$, $p < 0.001$), and a positive correlation with the variable of 'rivalry among competitors' ($r = 0.284$, $p < 0.001$).

As such, the better the relationships of an oil and gas service organisation with its key oil and gas clients, the more pleasant the business environment perceived by its senior executives; the more turbulent the rivalry among the competitors within the service sector of an organisation, the more unpleasant the business environment perceived by the senior executives of the organisation. Hence, the perceived environmental hostility is associated with relevant customers' and competitors' conditions.

To examine the perceived dynamism and associated environmental factors, Spearman's rho values are presented in Table 4.11. The results also indicate that the perceived dynamism is significantly associated with the oil and gas economic conditions. The perceived dynamism has a 0.215 significant correlation with the variable of 'changes in oil and gas E&P level' ($p = 0.034$); a 0.261 significant correlation with the variable of 'changes in rig counts' ($p = 0.011$); and a 0.174 marginally significant correlation with the variable of 'changes in well counts' ($p = 0.087$). As a result, if the unpredictability of changes in the oil and gas E&P levels, well counts and rig counts becomes higher, the business environment in which a service organisation operates is perceived to be more dynamic.

It is found that the perceived environmental dynamism has nothing to do with the variables of the other task environmental sectors, as their associations shown in the table are not significant. This means that, from year to year, even though details of changes in the regulatory, technological, customers', suppliers' and competitors' conditions cannot be forecast, each of the remaining task environmental sectors is not subject to a dynamic situation. The final conclusion is that, for the oil and gas service organisations in East Asian countries like China, Singapore and Malaysia, the perceived

Table 4.9 Correlations of perceived complexity and environmental factors

Spearman's rho		Complexity	Economics knowledge required	Technological level	Similarity of industrial products/ services
Complexity	Correlation coefficient	1.000	0.396**	0.386**	0.104
	Significance (2-tailed)	–	0.000	0.000	0.310
	N	98	98	98	98
Economics knowledge required	Correlation coefficient	0.396**	1.000	0.390**	−0.011
	Significance (2-tailed)	0.000	–	0.000	0.916
	N	98	98	98	98
Technological level	Correlation coefficient	0.386**	0.390**	1.000	−0.034
	Significance (2-tailed)	0.000	0.000	–	0.741
	N	98	98	98	98
Similarity of industrial products/ services	Correlation coefficient	0.104	−0.011	−0.034	1.000
	Significance (2-tailed)	0.310	0.916	0.741	–
	N	98	98	98	98
Government regulations legislation and policies	Correlation coefficient	0.063	0.099	0.046	0.075
	Significance (2-tailed)	0.536	0.333	0.649	0.464
	N	98	98	98	98
Number of oil and gas clients	Correlation coefficient	−0.130	−0.060	0.029	−0.005
	Significance (2-tailed)	0.201	0.555	0.779	0.959
	N	98	98	98	98
Needs and preferences of oil and gas clients	Correlation coefficient	0.226*	0.240*	0.188	0.282**
	Significance (2-tailed)	0.026	0.017	0.064	0.005
	N	98	98	98	98
Number of suppliers	Correlation coefficient	0.068	0.310**	0.251*	0.013
	Significance (2-tailed)	0.507	0.002	0.013	0.900
	N	97	97	97	97
Similarity of supply conditions by suppliers	Correlation coefficient	0.089	0.034	0.061	0.081
	Significance (2-tailed)	0.383	0.738	0.555	0.430
	N	97	97	97	97
Number of firms within the industry sector	Correlation coefficient	0.140	0.240*	−0.039	0.273**
	Significance (2-tailed)	0.169	0.017	0.701	0.007
	N	98	98	98	98
Scope of firms within the industry sector	Correlation coefficient	0.085	−0.044	0.109	−0.246*
	Significance (2-tailed)	0.407	0.669	0.284	0.015
	N	98	98	98	98

*Correlation is significant at the 0.05 level (2-tailed).
**Correlation is significant at the 0.01 level (2-tailed).

Government regulations, legislation and policies	Number of oil and gas clients	Needs and preferences of oil and gas clients	Number of suppliers	Similarity of supply conditions by suppliers	Number of firms within the industry sector	Scope of firms within the industry sector
0.063	−0.130	0.226*	0.068	0.089	0.140	0.085
0.536	0.201	0.026	0.507	0.383	0.169	0.407
98	98	98	97	97	98	98
0.099	−0.060	0.240*	0.310**	0.034	0.240*	−0.044
0.333	0.555	0.017	0.002	0.738	0.017	0.669
98	98	98	97	97	98	98
0.046	0.029	0.188	0.251*	0.061	−0.039	0.109
0.649	0.779	0.064	0.013	0.555	0.701	0.284
98	98	98	97	97	98	98
0.075	−0.005	0.282**	0.013	0.081	0.273**	−0.246*
0.464	0.959	0.005	0.900	0.430	0.007	0.015
98	98	98	97	97	98	98
1.000	−0.059	0.124	0.046	0.114	0.034	−0.037
–	0.562	0.224	0.657	0.268	0.739	0.715
98	98	98	97	97	98	98
−0.059	1.000	0.018	0.193	−0.006	0.076	−0.084
0.562	–	0.858	0.058	0.953	0.459	0.411
98	98	98	97	97	98	98
0.124	0.018	1.000	−0.006	−0.008	0.164	−0.071
0.224	0.858	–	0.956	0.940	0.106	0.486
98	98	98	97	97	98	98
0.046	0.193	−0.006	1.000	0.306**	0.346**	0.028
0.657	0.058	0.956	–	0.002	0.001	0.788
97	97	97	97	97	97	97
0.114	−0.006	−0.008	0.306**	1.000	0.227*	−0.073
0.268	0.953	0.940	0.002	–	0.025	0.475
97	97	97	97	97	97	97
0.034	0.076	0.164	0.346**	0.227*	1.000	0.059
0.739	0.459	0.106	0.001	0.025	–	0.566
98	98	98	97	97	98	98
−0.037	−0.084	−0.071	0.028	−0.073	0.059	1.000
0.715	0.411	0.486	0.788	0.475	0.566	–
98	98	98	97	97	98	98

Table 4.10 Perceived hostility and environmental factors

Spearman's rho		Hostility	Size of market demand	National government regulations and legislation	Local government policies	Relationship with government
Hostility	Correlation coefficient	1.000	0.107	0.142	−0.011	0.026
	Significance (2-tailed)	–	0.292	0.164	0.916	0.797
	N		98	98	98	98
Size of market demand	Correlation coefficient	0.107	1.000	0.237*	0.319**	0.281**
	Significance (2-tailed)	0.292	–	0.019	0.001	0.005
	N	98		98	98	98
National government regulations and legislation	Correlation coefficient	0.142	0.237*	1.000	0.526**	0.189
	Significance (2-tailed)	0.164	0.019	–	0.000	0.063
	N	98	98		98	98
Local government policies	Correlation coefficient	−0.011	0.319**	0.526**	1.000	0.370**
	Significance (2-tailed)	0.916	0.001	0.000	–	0.000
	N	98	98	98		98
Relationship with government	Correlation coefficient	0.026	0.281**	0.189	0.370**	1.000
	Significance (2-tailed)	0.797	0.005	0.063	0.000	–
	N	98	98	98	98	
Levels of key customer switching to competitor	Correlation coefficient	0.019	−0.085	0.147	0.173	0.247*
	Significance (2-tailed)	0.851	0.407	0.150	0.089	0.014
	N	98	98	98	98	98
Relationship with customers	Correlation coefficient	0.354**	0.443**	0.147	0.147	0.303**
	Significance (2-tailed)	0.000	0.000	0.148	0.147	0.002
	N	98	98	98	98	98
Access to suppliers for materials or goods	Correlation coefficient	0.173	0.148	0.156	0.082	0.057
	Significance (2-tailed)	0.092	0.151	0.130	0.429	0.578
	N	96	96	96	96	96
Relationship with suppliers	Correlation coefficient	0.105	0.291**	−0.026	0.064	0.197
	Significance (2-tailed)	0.308	0.004	0.799	0.535	0.054
	N	96	96	96	96	96
Entry barriers	Correlation coefficient	0.095	0.172	0.187	0.250*	0.129
	Significance (2-tailed)	0.354	0.089	0.066	0.013	0.207
	N	98	98	98	98	98
Rivalry among competitors	Correlation coefficient	0.284*	−0.092	0.181	0.094	−0.029
	Significance (2-tailed)	0.005	0.367	0.074	0.359	0.774
	N	98	98	98	98	98
Competitive actions	Correlation coefficient	0.139	−0.027	0.199*	0.192	0.065
	Significance (2-tailed)	0.171	0.789	0.050	0.059	0.527
	N	98	98	98	98	98
Relationship with competitors	Correlation coefficient	0.149	0.288**	0.080	0.057	0.159
	Significance (2-tailed)	0.143	0.004	0.431	0.575	0.118
	N	98	98	98	98	98

*Correlation is significant at the 0.01 level (2-tailed).
**Correlation is significant at the 0.05 level (2-tailed).

Levels of key customers switching to competitors	Relationship with customers	Access to suppliers for materials or goods	Relationship with suppliers	Entry barries	Rivalry among competitors	Competitive actions	Relationship with competitors
0.019	0.354**	0.173	0.105	0.095	0.284**	0.139	0.149
0.851	0.000	0.092	0.308	0.354	0.005	0.171	0.143
98	98	96	96	98	98	98	98
−0.085	0.443**	0.148	0.291**	0.172	−0.092	−0.027	0.288**
0.407	0.000	0.151	0.004	0.089	0.367	0.789	0.004
98	98	96	96	98	98	98	98
0.147	0.147	0.156	−0.026	0.187	0.181	0.199*	0.080
0.150	0.148	0.130	0.799	0.066	0.074	0.050	0.431
98	98	96	96	98	98	98	98
0.173	0.147	0.082	0.064	0.250*	0.094	0.192	0.057
0.089	0.147	0.429	0.535	0.013	0.359	0.059	0.575
98	98	96	96	98	98	98	98
0.247*	0.303**	0.057	0.197	0.129	−0.029	0.065	0.159
0.014	0.002	0.578	0.054	0.207	0.774	0.527	0.118
98	98	96	96	98	98	98	98
1.000	0.023	−0.060	−0.050	−0.055	0.286**	0.083	0.039
–	0.820	0.564	0.630	0.592	0.004	0.419	0.704
98	98	96	96	98	98	98	98
0.023	1.000	0.038	0.441**	0.082	−055	0.046	0.226*
0.820	–	0.713	0.000	0.423	0.590	0.652	0.026
98	98	96	96	98	98	98	98
−0.060	0.038	1.000	0.268**	0.206*	0.265**	0.188	0.169
0.564	0.713	–	0.008	0.044	0.009	0.066	0.100
96	96	96	96	96	96	96	96
−0.050	0.441**	268**	1.000	089	−0.003	−0.048	0.277**
0.630	0.000	0.008	–	0.389	0.975	0.645	0.006
96	96	96	96	96	96	96	96
−0.055	0.082	0.206*	0.089	1.000	0.158	0.198	0.162
0.592	0.423	0.044	0.389	–	0.120	0.051	0.112
98	98	96	96	98	98	98	98
0.286**	−0.055	0.265**	−0.003	0.158	1.000	0.505**	0.322**
0.004	0.590	0.009	0.975	0.120	–	0.000	0.001
98	98	96	96	98	98	98	98
0.083	0.046	0.188	−0.048	0.198	0.505**	1.000	0.434**
0.419	0.652	0.066	0.645	0.051	0.000	–	0.000
98	98	96	96	98	98	98	98
0.039	0.226*	0.169	0.277**	0.162	0.322**	0.434**	1.000
0.704	0.026	0.100	0.006	0.112	0.001	0.000	–
98	98	96	96	98	98	98	98

Table 4.11　Perceived dynamism and environmental factors

Spearman's rho		Perceived dynamism
Perceive dynamism	Correlation coefficient	1.000
	Significance (2-tailed)	–
	N	98
Customer demand	Correlation coefficient	–0.036
for existing	Significance (2-tailed)	0.721
products/services	N	98
Customer demand for	Correlation coefficient	–0.137
new products/services	Significance (2-tailed)	0.180
	N	98
Customer demand for	Correlation coefficient	–0.149
higher quality or more	Significance (2-tailed)	0.142
service	N	98
Customer preference for	Correlation coefficient	0.059
lower prices	Significance (2-tailed)	0.562
	N	98
Changes in competitive	Correlation coefficient	–0.009
price	Significance (2-tailed)	0.933
	N	98
Competitors' quality	Correlation coefficient	0.017
improvement	Significance (2-tailed)	0.867
	N	98
Competitors'	Correlation coefficient	0.143
introduction of new	Significance (2-tailed)	0.160
products/services	N	98
Suppliers' rising prices	Correlation coefficient	–0.070
	Significance (2-tailed)	0.495
	N	97
Suppliers' quality	Correlation coefficient	0.086
reduction	Significance (2-tailed)	0.401
	N	97
Suppliers' introduction	Correlation coefficient	–0.011
of new materials or	Significance (2-tailed)	0.913
standard products	N	97
Changes in oil and gas	Correlation coefficient	0.215*
E&P level	Significance (2-tailed)	0.034
	N	98
Changes in well counts	Correlation coefficient	0.174[a]
	Significance (2-tailed)	0.087
	N	97

Table 4.11 Continued

Spearman's rho		Perceived dynamism
Changes in rig counts	Correlation coefficient	0.261*
	Significance (2-tailed)	0.011
	N	95
Technological changes	Correlation coefficient	–0.011
	Significance (2-tailed)	0.918
	N	97
Rate of technological diffusion	Correlation coefficient	0.150
	Significance (2-tailed)	0.142
	N	97
Changes in national regulations and legislation	Correlation coefficient	0.016
	Significance (2-tailed)	0.872
	N	98
Changes in local government policies	Correlation coefficient	–0.017
	Significance (2-tailed)	0.867
	N	98

[a] Correlation is marginally significant at the 0.10 level (2-tailed).
*Correlation is significant at the 0.05 level (2-tailed).

environmental dynamism is associated with the unpredictability of the oil economic conditions.

Based upon the above analysis, the level of perceived environmental complexity, dynamism and hostility is associated with the influences of the task environmental factors for the oil and gas service sector in East Asia. Hence, Proposition 6 is supported by the research results.

4.5 Cross-national comparisons

4.5.1 Conditions in China, Singapore and Malaysia

In order to emphasise the differences and similarities pictorially among the items, Figure 4.8 shows the bar charts of the variables of the environmental dimensions split according to the country where the service organisations were located.

First, the distribution of the perceived environmental uncertainty scores appears to be normal. This result explains the observations made in Section 4.3.1 that the overall uncertainty score is normally distributed (Figure 4.6). It is observed that the perceived uncertainty in China was the lowest ($M = 3.6$), whereas in Singapore, the managerial perception on the degree of environmental uncertainty was

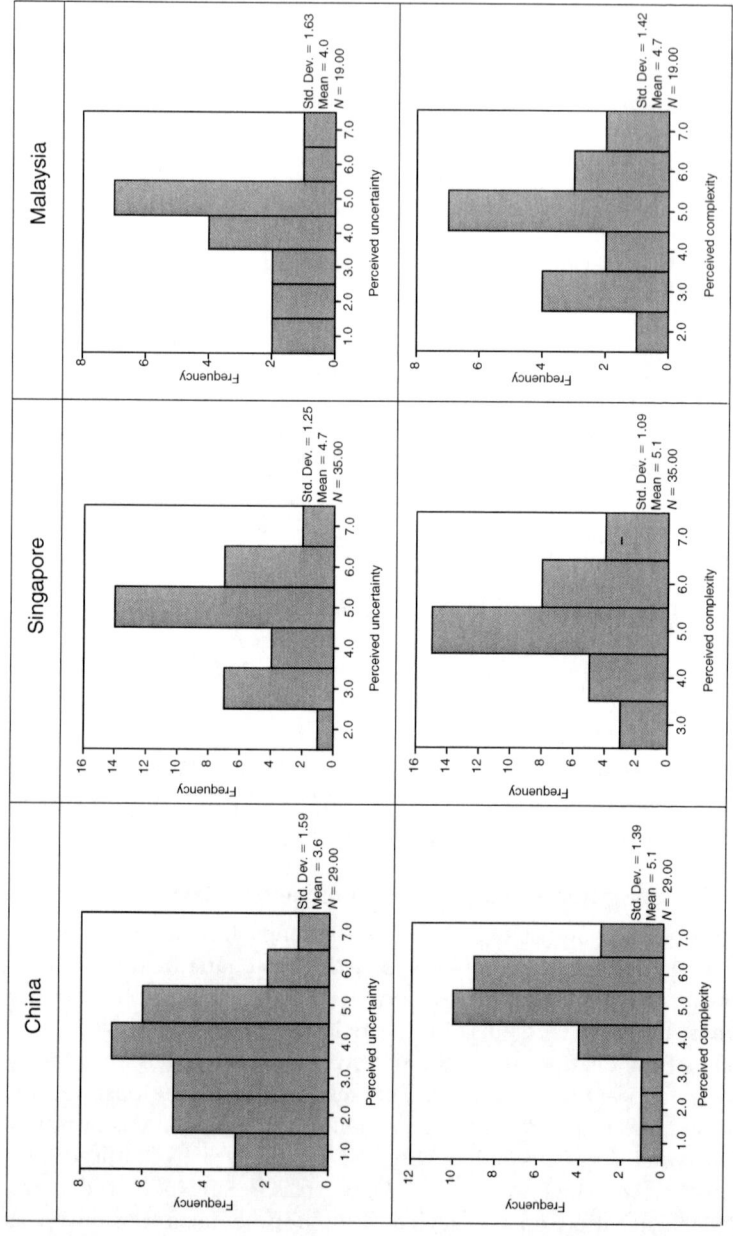

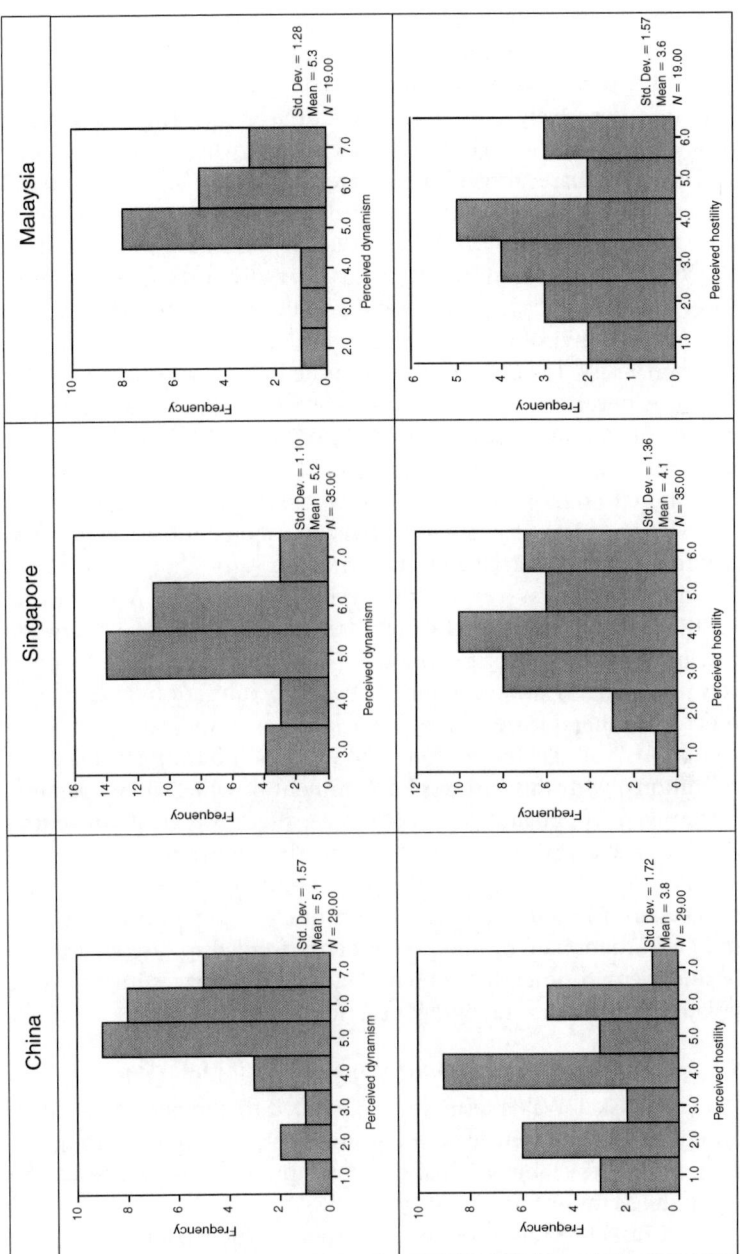

Figure 4.8 Perceived business environment in China, Singapore and Malaysia

the highest (M = 4.7). The Malaysian organisations' executives had a neutral view on environmental uncertainty (M = 4).

For the perceived environmental complexity scores, the distribution is negatively skewed in the China group (there is a larger concentration at the higher end of the scores), whereas the Singapore and Malaysia-based organisations are normally distributed around a mean of 5.1 and 4.7 respectively. Therefore, the overall negative skew observed in Section 4.3.1 is due to the mixture of countries (the Chinese organisations affect the normally distributed scores of Singapore and Malaysia). The results indicate that in each country, the business environment is perceived to be complex.

Nevertheless, for perceived environmental dynamism, the distribution is negatively skewed in both the China and Malaysia groups yet the Singapore organisations are normally distributed around a mean of 5.2. Hence, the China and Malaysia-based organisations contribute to the overall negative skew observed earlier in Figure 4.6 (see Section 4.3.1). It is observed that, in each country, the business environment was perceived to be dynamic.

When looking at the distribution of perceived environmental hostility scores, all the distributions are bimodal (i.e. there are two peaks indicative of two modes). It seems that regardless of the country, there is always a split between organisations' executives: they are either pleased (one mode at 2 or 3 indicating pleasant conditions) or displeased (one mode at 6 indicating unpleasant conditions) with the business environment in which they operate.

However, for Singapore and Malaysia, there is a great concentration of executives' perceptions around the higher mode (the peak is taller) at 4, indicating the neutral point between pleasant and unpleasant. In each of the three countries, a common feature shows that less than half of the respondents were displeased with the business environment in which they operate, suggesting they are favourable places for oil and gas service businesses.

4.5.2   Comparisons of China, Singapore and Malaysia

Overall Kruskal–Wallis tests were conducted to compare the median differences of the perceived environmental conditions among the three countries (Table 4.12) and follow-up tests were carried out by using the Mann–Whitney U-test.

The Kruskal–Wallis tests indicate that in each country, there is no significant difference in the medians of dynamism χ^2 = 0.259

Table 4.12 Perceived business environment in China, Singapore and Malaysia

Ranks

	Country where based	N	Mean rank
Perceived	China	31	46.95
dynamism	Singapore	39	44.45
	Malaysia	21	47.48
	Total	91	
Perceived	China	31	49.81
complexity	Singapore	39	47.40
	Malaysia	21	37.79
	Total	91	
Perceived	China	31	43.40
hostility	Singapore	39	49.28
	Malaysia	21	43.74
	Total	91	
Perceived	China	31	55.16
uncertainty	Singapore	39	40.06
	Malaysia	21	43.50
	Total	91	

Test statistics[a,b]

	Perceived dynamism	Perceived complexity	Perceived hostility	Perceived uncertainty
Chi-square	0.259	2.991	1.098	6.991
df	2	2	2	2
Asymptotic significance	0.878	0.224	0.578	0.030

[a] Kruskal–Wallis test.
[b] Grouping variable: Country where based.

(df = 2), $p = 0.878$, complexity $\chi^2 = 2.991$ (df = 2), $p = 0.224$ or hostility $\chi^2 = 1.098$ (df = 2), $p = 0.578$. In contrast, the same tests show a significant difference in the medians of the perceived uncertainty $\chi^2 = 6.991$ (df = 2), $p = 0.030$ in these three countries.

There is no evidence to suggest that managerial perceptions on environmental complexity, dynamism and hostility in the three selected countries are dissimilar. The results fail to reject the null hypothesis that managerial perceptions on environmental complexity, dynamism and hostility in the three selected countries are dissimilar. In order to prove similarity, bigger samples are needed. Nonetheless, the observed results may indicate that executives in

Table 4.13   Perceived uncertainty in China and Singapore

Ranks

	Country where based	N	Mean rank	Sum of ranks
Perceived uncertainty	China	31	27.73	859.50
	Singapore	39	41.68	1625.50
	Total	70		

Test statistics[a]

	Perceived uncertainty
Mann–Whitney *U*	363.500
Wilcoxon *W*	859.500
Z	−2.911
Asymptotic significance (2-tailed)	0.004

[a] Grouping variable: Country where based.

China, Singapore and Malaysia shared similar views on the environmental complexity, dynamism and hostility.

Because the overall test for the perceived environmental uncertainty is significant, pairwise comparisons among the three groups of countries were conducted. The selected SPSS results with the Mann–Whitney test are presented in Table 4.13.

Only the comparison between China and Singapore is significant. It suggests that senior executives in Singapore had rather different views from those in China: the degree of perceived uncertainty in Singapore was much higher than that in China. However, no evidence is given to show the views in Malaysia were dissimilar to those in Singapore or China. As such, executives in Malaysia might share similar opinions to those either in Singapore or China.

4.6   Summary

Managerial perceptions on environmental influences differed significantly. The impact of economic, customers' and competitors' factors was strong, and especially the customers' factor having the strongest impact. In contrast, the suppliers' factor had the least impact on the businesses of the participating organisations.

Research results have proved the assumption that the higher the degree of the perceived environmental dynamism and hostility, the higher the degree of perceived uncertainty. However, the results do not support the assumption that there is a relationship between perceived complexity and perceived uncertainty.

The results suggest that the more complicated the government regulations, legislations and policies in the region where a service organisation operates in East Asia, or the more different the supply conditions (e.g. price, quality, speed or service) provided by the organisation's suppliers, the higher the degree of the perceived environmental uncertainty.

The perceived environmental complexity was found to be associated with the level of technologies existing in the industry, the customer-related factors and oil economic conditions.

The perceived environmental dynamism is associated with the unpredictability of the oil economic conditions. If the unpredictability of changes in the oil and gas E&P levels, well counts and rig counts becomes higher, the business environment in which a service organisation operates is perceived to be more dynamic.

The perceived environmental hostility is associated with relevant customers' and competitors' conditions. It means that, the better the relationships of an oil and gas service organisation with its key oil and gas clients, the more pleasant the business environment perceived by its senior executives; the more turbulent the rivalry among the competitors within the service sector of an organisation, the more unpleasant the business environment perceived by the senior executives of the organisation.

Overall, the business environment for the service sector in East Asia was found to be slightly uncertain, with the lowest level in China and the highest in Singapore. China appeared to have a steady business environment and the situation of environmental uncertainty in Malaysia is neutral. Only in Singapore is the managerial perception of the business environment uncertain.

Although the business environment is complex and dynamic in each country, the environmental conditions are predictable and attractive to oil and gas service companies. The majority of the participating oil and gas service organisations in East Asia had amicable relationships with their competitors, although competition could be very strong.

5
Business Strategies in China, Singapore and Malaysia

This chapter examines and categorises different business strategies deployed by the participating service organisations. It reveals competitive strategies and strategic positions for the service companies operating in the East Asian market. The chapter also compares differences and similarities of strategic directions and organisational performance across the three selected East Asian countries: China, Singapore and Malaysia.

5.1 Business strategies adopted by service companies

This section seeks to investigate Proposition 7 that, for oil and gas service organisations operating in East Asian countries like China, Singapore and Malaysia, managerial perceptions of their business strategies will be different. Based on the theoretical frameworks generated in Chapter 3, the participating organisations can be categorised into different strategic groups.

5.1.1 Generic business strategies employed by oil and gas service firms

The present study allocates strategic orientations for businesses based on Miles and Snow's (1978) theory and an approach developed by Parnell et al. (2000). A business was categorised as pursuing the strategy reflected by the highest number of its responses and was firstly assigned into the categories of defender, prospector and reactor.

Table 5.1   Strategic orientations by service companies

	Frequency	%
Defender	4	4.1
Prospector	5	5.1
Analyser	59	60.2
Balancer	17	17.3
Reactor	13	13.3
Total	98	100.0

When there was a tie between defender and prospector strategies, the business was assigned as an analyser. When there was a tie among three non-reactor strategies, the business was classified as a balancer. To distinguish the characteristics of a balancer from those of an analyser, the business should indicate its capability of 'adapting to' and 'creating' changes or innovation simultaneously. Finally, when there was a tie between the reactor and one or two other strategies, the business was still classified as a reactor.

By carrying out the above procedure case by case, the 98 participating oil and gas service organisations were assigned as 17 balancers, 59 analysers, 5 prospectors, 4 defenders and 13 reactors (Table 5.1). A large majority of the participating organisations had developed well-defined strategies guiding their business practice in East Asia (86.6 per cent). Most of the participating organisations were analysers (60.2 per cent) and the fewest organisations were defenders (4.1 per cent).

5.1.2   Trends of competitive strategies

An approach categorising competitive strategies was developed based upon Porter's typology.

If firms competed on the basis of highest quality or differentiated themselves from competitors in the industry, they were assigned to the category of differentiation organisations; if firms competed based on the lowest cost in the industry, they were categorised as low-cost organisations. If firms pursued both the lowest possible cost and the highest possible quality or a unique feature, and were able to compete simultaneously based on both the lowest cost and differentiated business, they were classed as hybrid organisations. If firms pursued one competitive strategy and were also able to compete

Table 5.2   Competitive strategies of the service organisations

	Frequency	%
Low cost	5	5.1
Differentiation	48	49.0
Hybrid	36	36.7
No-purpose	9	9.2
Total	98	100.0

based on combined competitive advantages, they were still assigned to a hybrid strategy.

In contrast, if firms sought to attain both cost leadership and differentiation yet failed to achieve a strategy in either direction, they were assigned to the category of no-purpose organisations. If firms pursued one type of strategy but in fact appeared to have other types of competitive advantage, these firms were also no-purpose organisations. If firms pursued a hybrid strategy while failing to achieve either low cost or differentiation, they were assigned to an associated strategy – either one of the two basic competitive strategies.

By carrying out the above procedures, the 98 participant oil and gas service organisations were classified into 48 differentiation, 36 hybrid, 5 low-cost and 9 no-purpose organisations (Table 5.2). Overall, nearly half of the participating firms deployed a differentiation-oriented competitive strategy (49 per cent) and a minority of firms pursued a low-cost strategy (5.1 per cent), which was obviously not a preferred option selected by senior management of oil and gas service companies operating in East Asia. The result supports my contention that one of the typical attributes of the oil and gas service industry is differentiation.

5.1.3   Strategic competitive positions

A scatter graph was used to plot the profile of price level and perceived customer added value (PCAV). In order to obtain a value of PCAV, the average value of five variables (i.e. products or service quality valued by customers, reliability of technology, safety performance of products or services, speed of response to clients' requirements and price that customers are willing to pay) was adopted. In order to

Table 5.3   Competitive positions of the service companies

Competitive positions	Competitive strategy	Frequency	%
Low (or moderate) value competitive price	Low cost	5	5.1
High value competitive (lower than moderate) price	Hybrid	7	7.1
High value moderate price	Hybrid	28	28.6
High value premium price	Differentiation	33	33.7
Uncompetitive value and price	No-purpose	25	25.5
Total		98	100.0

eliminate a self-report bias, the mean of the perceived PCAV was used as the central tendency in the service industrial sector.

By carrying out the above procedure, it was possible to categorise five clusters of strategic competitive positions for the participating organisations. As defined in Section 3.4.3, the five positions are: high value premium (i.e. above the moderate level) price, high value moderate price, high value competitive (i.e. below the moderate level) price, low or moderate value low price, and uncompetitive value and price.

Using frequency statistics, the results presented in Table 5.3 show that, of the 98 participating organisations, 33 were in the high value premium price competitive position, 28 high value moderate price, 7 high value competitive (lower than moderate price), 5 low (or moderate) value competitive price, and 25 uncompetitive value and price.

In order to simplify the results, four clusters of competitive positions were obtained (Figure 5.1). Position 1 indicates the 'no frills' (Johnson and Scholes, 1999) group, comprising organisations which were likely to target the low-cost low-added-value sector. Such organisations believed that they would not be defended by their competitors. The price of their products or services was lower than the moderate industrial level and brought limited added value prospects.

For organisations at Position 2, the quality, reliability and safety performance of their products or services and their speed of response to clients' requirements were higher than the moderate industrial levels. They could sell their products or services at a relatively competitive price. They combined both price and value advantages.

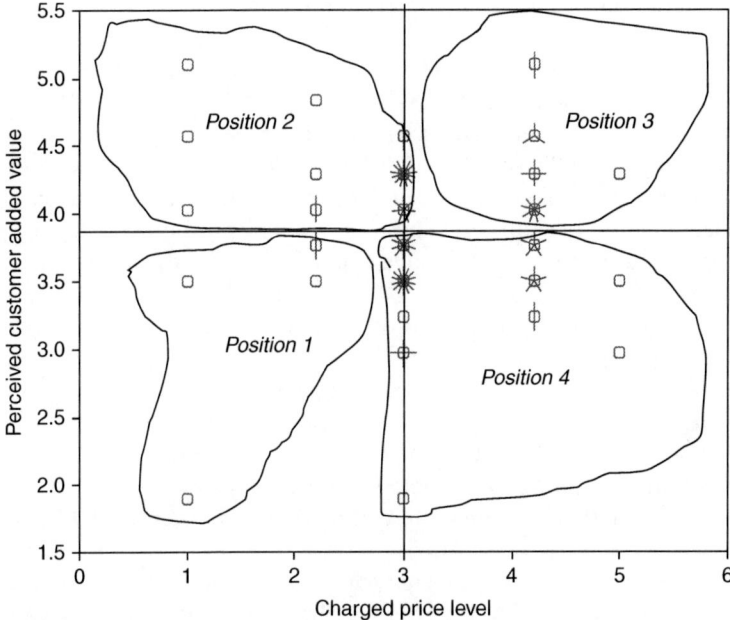

Figure 5.1 Strategic groups of the service companies*
Note: in Figure 5.1, the more cases, the more petals.

In Position 3, organisations sought to establish a premium price by providing products or services which had unique features within their particular market segments.

Position 4 indicates the organisations which employed strategies 'destined for ultimate failure' (Johnson and Scholes, 1999) as they had neither value nor price advantages for successful competition in the marketplace.

In conclusion of the above three sections, Proposition 7 pertaining to different business strategies is supported.

5.2 Key competitive factors in the service industry

Competitive issues are related to long-term organisational success. This section intends to explore further that, for oil and gas service organisations operating in the three selected countries, there will be a relationship between strategic performance and key industrial competitive factors.

Using the Spearman correlation and rho (r) values, these factors were examined in association with strategic performance. As the prediction was not directional, a two-tailed test was selected.

A matrix was displayed giving the correlation coefficient between strategic performance and one of the competitive factors. It was found that there is a significantly positive relationship between strategic performance and each of the five competitive features (Table 5.4):

- products' or services' quality valued by customers, $r = 0.25$, $p = 0.022$;
- reliability of technologies of products or services, $r = 0.26$, $p = 0.018$;
- safety performance of products and services, $r = 0.22$, $p = 0.044$;
- speed of response to customers, $r = 0.25$, $p = 0.021$;
- and price that customers are willing to pay, $r = 0.24$, $p = 0.025$.

As all correlation coefficient values are positive, it can be concluded that as the level of one of the above competitive advantages improves, there is a corresponding improvement in strategic performance. Hence, the assumption that there is a relationship between strategic performance and the defined competitive factors was supported.

5.3   Strategic performance

The data were measured by using an interval scale. The box plot (Figure 5.2) for strategic performance looks moderately symmetrical as the box is almost in the middle of the whiskers and the median is only slightly above the middle of the box. This suggests that these data are only very slightly skewed and therefore the mean may be employed as the appropriate measure of central tendency (Kerr et al., 2002).

Overall, the performance of the participating organisations had improved in each of the 20 defined performing areas during the five years (1997–2001) that were reviewed.

Using histogram descriptive statistics, Figure 5.3 presents the distribution of strategic performance scores. The mean of the strategic

Table 5.4 Correlations of key competitive factors and strategic performance*

			Correlations				
			Quality valued by customers	Reliability of (products/services) technology	Safety performance of products/services	Speed of response to clients' requirements	Price that customers are willing to pay	Strategic performance
Spearman's rho	Quality valued by customers	Correlation coefficient	1.000	0.496**	0.567**	0.209**	0.264*	0.246*
		Significance (2-tailed)	–	0.000	0.000	0.007	0.015	0.022
		N	87	85	85	85	85	87
	Reliability of (products/services) technology	Correlation coefficient	0.496**	1.000	0.441**	0.013	0.150	0.256*
		Significance (2-tailed)	0.000	–	0.000	0.906	0.171	0.018
		N	85	85	85	85	85	85
	Safety performance of products/services	Correlation coefficient	0.567**	0.441**	1.000	0.434**	0.266*	0.219*
		Significance (2-tailed)	0.000	0.000	–	0.000	0.014	0.044
		N	85	85	85	85	85	85
	Speed of response to clients' requirements	Correlation coefficient	0.290**	0.13	0.434**	1.000	0.209	0.250*
		Significance (2-tailed)	0.007	0.906	0.000	–	0.055	0.021
		N	85	85	85	85	85	85
	Price that customers are willing to pay	Correlation coefficient	0.264**	0.150	0.266*	0.209	1.000	0.244*
		Significance (2-tailed)	0.015	0.171	0.014	0.055	–	0.025
		N	85	85	85	85	85	85
	Strategic performance	Correlation coefficient	0.246*	0.256*	0.219*	0.250*	0.244*	1.000
		Significance (2-tailed)	0.022	0.018	0.044	0.021	0.025	–
		N	87	85	85	85	85	87

*Correlation is significant at the 0.05 level (2-tailed).
**Correlation is significant at the 0.01 level (2-tailed).
*Note: Excluding 11 organisations that had been in business for less than 5 years and whose functional managers responded to the survey.

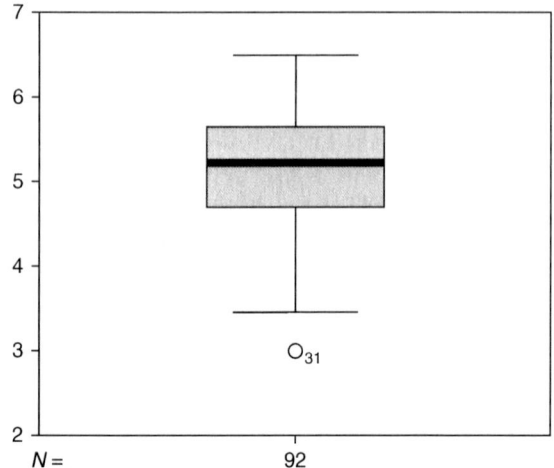

Figure 5.2 Strategic performance of the service companies
Note: 92 cases with no fewer than five years of business were selected.

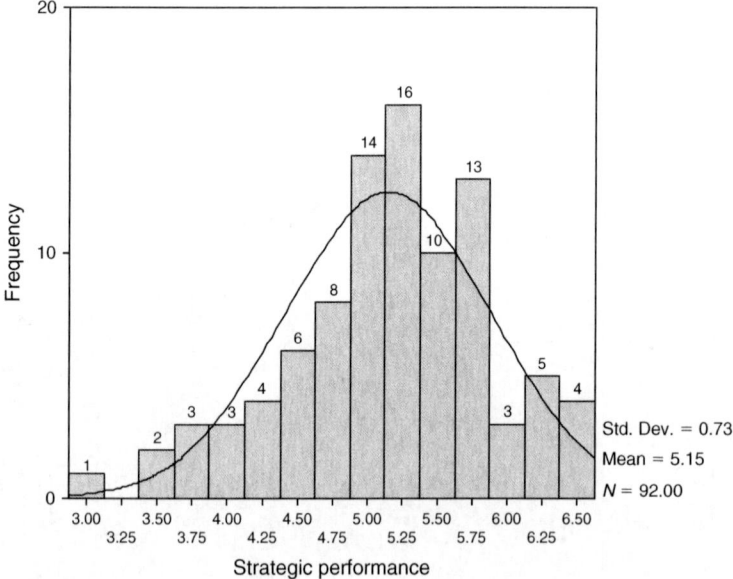

Figure 5.3 Assessing individual strategic performance

performance of the 92 participating organisations with no fewer than five years of business is 5.15, indicating that more than half (the majority) of them had improved their strategic performance over a period of the five years reviewed (i.e. 1997–2001). This reflects the favourable status of oil and gas service organisations operating in China, Singapore and Malaysia.

5.4 Cross-national comparisons of strategic options

5.4.1 Business strategies employed in China, Singapore and Malaysia

The following clustered bar chart (Figure 5.4) shows the trends of strategic orientations in the three selected countries. A cross-tabulation table is attached to Figure 5.4 to demonstrate further details of the results.

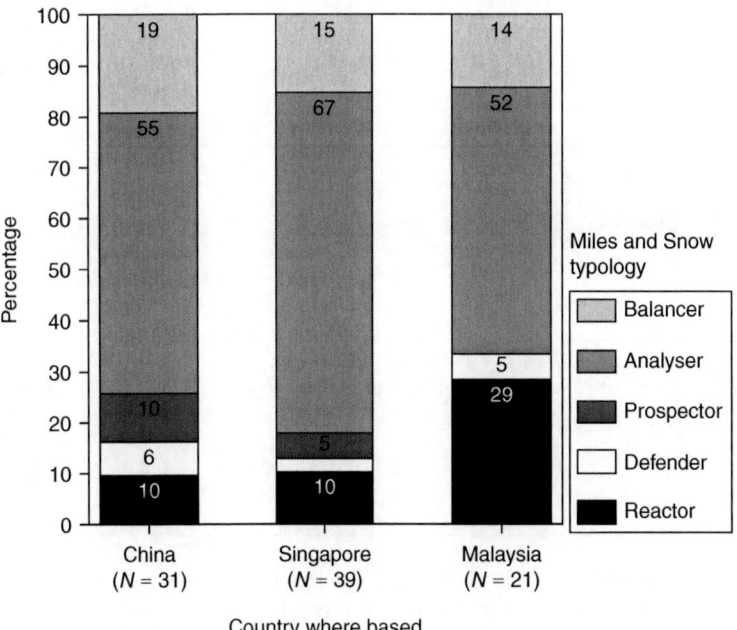

Figure 5.4 Cross-national comparison of business strategies

Miles and Snow – country where based cross-tabulation

			Country where based			Total
			China	Singapore	Malaysia	
Miles and Snow	Reactor	Count	3	4	6	13
		% within Miles and Snow	23.1	30.8	46.2	100.0
		% within country where based	9.7	10.3	28.6	14.3
	Defender	Count	2	1	1	4
		% within Miles and Snow	50.0	25.0	25.0	100.0
		% within country where based	6.5	2.6	4.8	4.4
	Prospector	Count	3	2		5
		% within Miles and Snow	60.0	40.0		100.0
		% within country where based	9.7	5.1		5.5
	Analyser or Balancer	Count	23	32	14	69
		% within Miles and Snow	33.3	46.4	20.3	100.0
		% within country where based	74.2	82.1	66.7	75.8
Total		Count	31	39	21	91
		% within Miles and Snow	34.1	42.9	23.1	100.0
		% within country where based	100.0	100.0	100.0	100.0

The similarity over the three countries is that most organisations were balancers or analysers with a few adopting a defender strategy in the country where they operate. The proportion of balancers and analysers was higher in Singapore (82.1 per cent) than in China (74.2 per cent) or Malaysia (66.7 per cent).

Overall, the majority of balancers and analysers were from Singapore. Of the 69 balancers and analysers, the proportions in China, Singapore and Malaysia were 33.3, 46.4 and 20.3 per cent respectively.

Out of four defenders, two were from China and one from Singapore and one from Malaysia. None of the five prospectors was from Malaysia, while three of those were China-based organisations and two were from Singapore.

The greatest proportion of 13 reactors was from Malaysia (46.2 per cent) and the fewest from China (23.1 per cent). The results suggest that Singapore organisations were more likely to conduct multiple strategies than the China and Malaysia-based organisations.

5.4.2 Competitive strategies in China, Singapore and Malaysia

Figure 5.5 compares the differences between competitive strategies applied in the three selected countries. In Singapore, most organisations emphasise creating a unique feature within the industry (64 per cent).

The majority of the China-based organisations emphasise differentiating themselves from competitors (45.2 per cent). However, in Malaysia, most of the 19 organisations were keen on a combined advantage such as pursuing a hybrid strategy (57.1 per cent).

Likewise, a two-way contingency table is presented to compare the respective competitive strategies used in China, Singapore and Malaysia (Table 5.5). The proportions of the service organisations which adopted a hybrid strategy towards China, Singapore and Malaysia were 28.6, 37.1 and 34.3 per cent respectively. For the category of a differentiation strategy, the associated proportions were 31.8, 56.8 and 11.4 per cent.

Most of the differentiation organisations were from Singapore. No low-cost organisations were found operating in Singapore. Five out of seven no-purpose organisations were from China, one from Singapore and one from Malaysia.

Additionally, in Singapore and Malaysia, the lowest percentage of competitive strategic type were the no-purpose organisations, with 2.6 and 4.8 per cent respectively. In China, however, the lowest percentage of competitive strategies was found to be low-cost organisations (6.5 per cent) rather than no-purpose organisations (16.1 per cent).

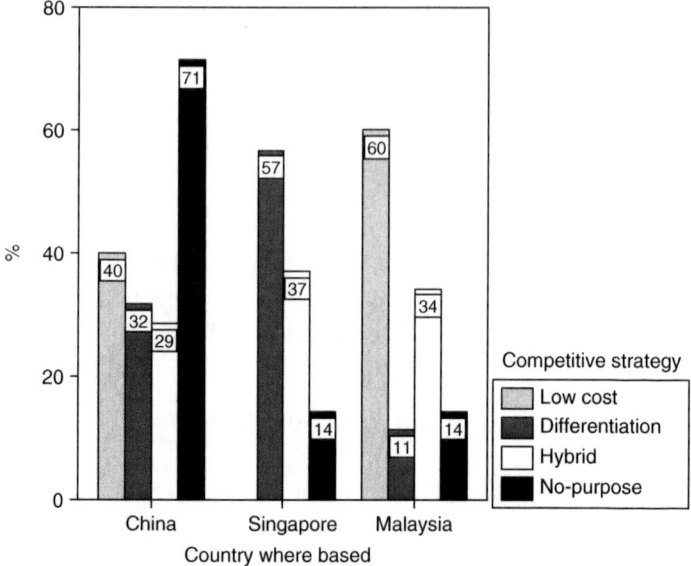

(a) Distributions in each country

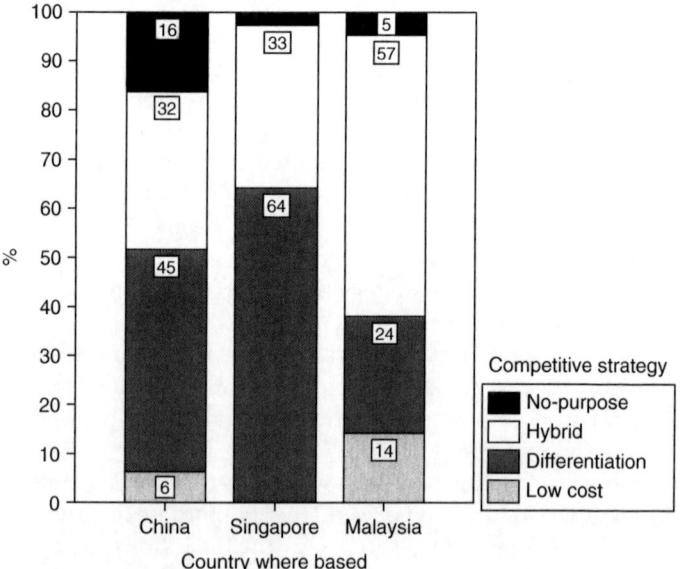

(b) Comparisons of individual country

Figure 5.5 Comparisons of competitive strategies applied in China, Singapore and Malaysia

Table 5.5 Competitive strategy by country where based*

			Country where based			Total
			China	Singapore	Malaysia	
Competitive strategy by Porter	Low cost	Count	2	0	3	5
		Expected count	1.7	2.1	1.2	5.0
		% within competitive strategy by Porter	40.0	0	60.0	100.0
		% within country where based	6.5	0	14.3	5.5
		% of Total	2.2	0	3.3	5.5
	Differentiation	Count	14	25	5	44
		Expected count	15.0	18.9	10.2	44.0
		% within competitive strategy by Porter	31.8	56.8	11.4	100.0
		% within country where based	45.2	64.1	23.8	48.4
		% of Total	15.4	27.5	5.5	48.4
	Hybrid	Count	10	13	12	35
		Expected count	11.9	15.0	8.1	35.0
		% within competitive strategy by Porter	28.6	37.1	34.3	100.0
		% within country where based	32.3	33.3	57.1	38.5
		% of Total	11.0	14.3	13.2	38.5
	No-purpose	Count	5	1	1	7
		Expected count	2.4	3.0	1.6	7.0
		% within competitive strategy by Porter	71.4	14.3	14.3	100.0
		% within country where based	16.1	2.6	4.8	7.7
		% of Total	5.5	1.1	1.1	7.7
Total		Count	31	39	21	91
		Expected count	31.0	39.0	21.0	91.0
		% within competitive strategy by Porter	34.1	42.9	23.1	100.0
		% within country where based	100.0	100.0	100.0	100.0
		% of Total	34.1	42.9	23.1	100.0

Note: As 50% of cells have an expected count less than 5, and the minimum expected count is 1.2, the resulting p-value for the overall chi-square test may not be trustworthy and hence the coefficient value of chi-square test is not applied.

In summary, the similarity in the three countries is that the majority of the organisations did endeavour to pursue a differentiation-oriented strategy. They sought to emphasise a unique feature of their businesses and differentiate themselves from others. Singaporean organisations were more likely to seek to achieve a differentiation character than the Chinese and Malaysian organisations.

5.4.3 Cross-national comparison of strategic positions

The same two-way contingency table analysis was used to evaluate the situation regarding competitive positions in the three selected countries (Table 5.6). The observed results are also presented in Figure 5.6.

Singaporean organisations appeared to have better strategic positions than the China and Malaysia-based organisations. Most of the high value premium organisations (61.3 per cent) and the majority of the high value competitive price organisations (40.6 per cent) were from Singapore. Most organisations following a strategy destined for ultimate failure with both uncompetitive price and value were from China (50.0 per cent). None of the four organisations with low or moderate value and competitive price was from Singapore, whereas three of them were from Malaysia and one from China.

In Singapore, the majority of the organisations fell into the category of high value premium price organisations (48.7 per cent). Most of the Malaysian organisations pursued a combined advantage of high value and competitive (low and moderate) price (52.4 per cent). For China, the organisations pursued mainly three types of strategic positions and the majority fell into the group of uncompetitive price and value (38.7 per cent).

In short, there was nothing in common in terms of competitive positions in these three countries. In this sense, the strategic approaches adopted by oil and gas service organisations for competing businesses in the region of East Asia varied greatly in their nature. Notably, organisations in Singapore appeared to be in a better competitive position compared with those in China and Malaysia.

Table 5.6 Strategic positions by country where based*

			Country where based			Total
			China	Singapore	Malaysia	
Strategy clock positions	Low (or moderate) value competitive price	Count	1	0	3	4
		Expected count	1.4	1.7	0.9	4.0
		% within strategy clock positions	25.0	0	75.0	100.0
		% within country where based	3.2	0	14.3	4.4
		% of Total	1.1	0	3.3	4.4
	High value competitive (low and moderate) price	Count	8	13	11	32
		Expected count	10.9	13.7	7.4	32.0
		% within strategy clock positions	25.0	(40.6)	34.4	100.0
		% within country where based	25.8	33.3	(52.4)	35.2
		% of Total	8.8	14.3	12.1	35.2
	High value premium price	Count	10	19	2	31
		Expected count	10.6	13.3	7.2	31.0
		% within strategy clock positions	32.3	(61.3)	6.5	100.0
		% within country where based	32.3	48.7	9.5	34.1
		% of Total	11.0	20.9	2.2	34.1
	Uncompetitive value and price	Count	12	7	5	24
		Expected count	8.2	10.3	5.5	24.0
		% within strategy clock positions	(50.0)	29.2	20.8	100.0
		% within country where based	38.7	17.9	23.8	26.4
		% of Total	13.2	7.7	5.5	26.4
Total		Count	31	39	21	91
		Expected count	31.0	39.0	21.0	91.0
		% within strategy clock positions	34.1	42.9	23.1	100.0
		% within country where based	100.0	100.0	100.0	100.0
		% of Total	34.1	42.9	23.1	100.0

*Note: As 25% of cells have an expected count less than 5, and the minimum expected count is 0.92, the resulting *p*-value for the overall chi-square test may not be trustworthy and hence the coefficient value of chi-square test is not applied.

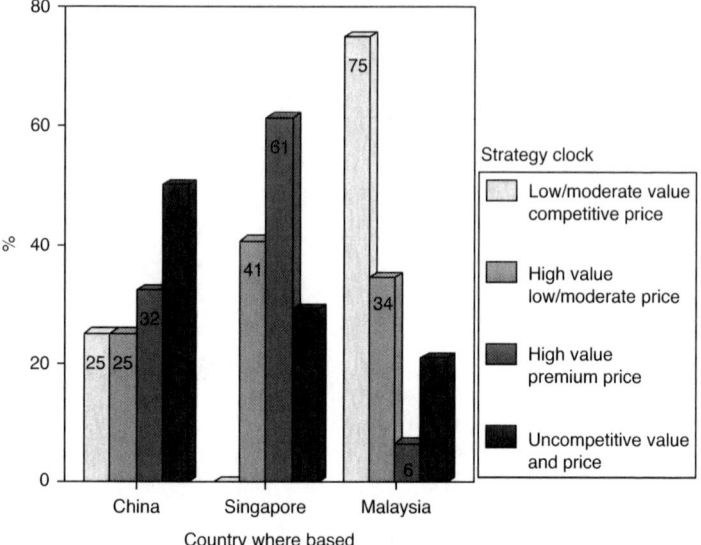

(a) Distributions in each country

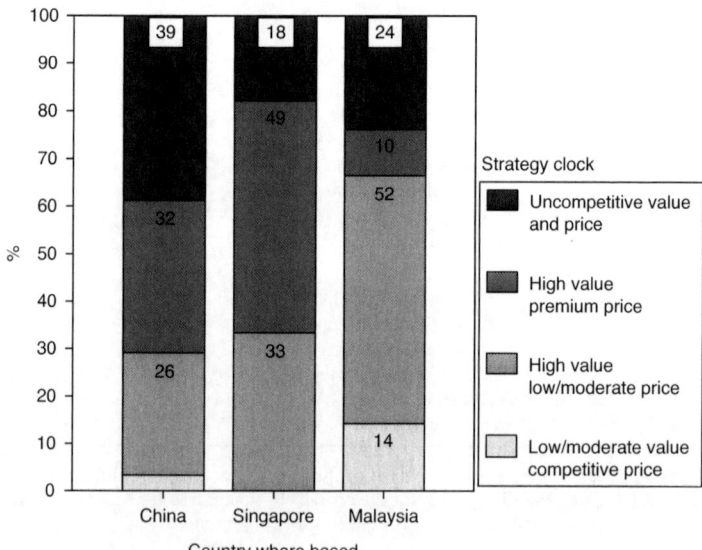

(b) Comparisons of individual country

Figure 5.6 Comparisons of strategic positions in China, Singapore and Malaysia

5.5   Cross-national analysis of strategic performance

An overall one-way analysis of variance (ANOVA) was applied for the assessment of whether the means on strategic performance were significantly different among the country groups. Where appropriate, post hoc multiple comparisons were conducted in an instance where the variances were assumed as equal.

5.5.1   An overall view of strategic performance

The resulting chart is shown in Figure 5.7. There was a marked difference between the 'become better' bar and 'become worse' or 'no change' bars in each of the country groups.

For the China-based organisations, most (79 per cent) had improved their strategic performance. A similar pattern is observed for the Singapore and Malaysia groups.

It is noted that the low performance (become worse) bar only appeared in the Singapore group. The no change bar for the China group is larger than both the Singapore and Malaysia groups. This indicates that a higher proportion of organisations in China than

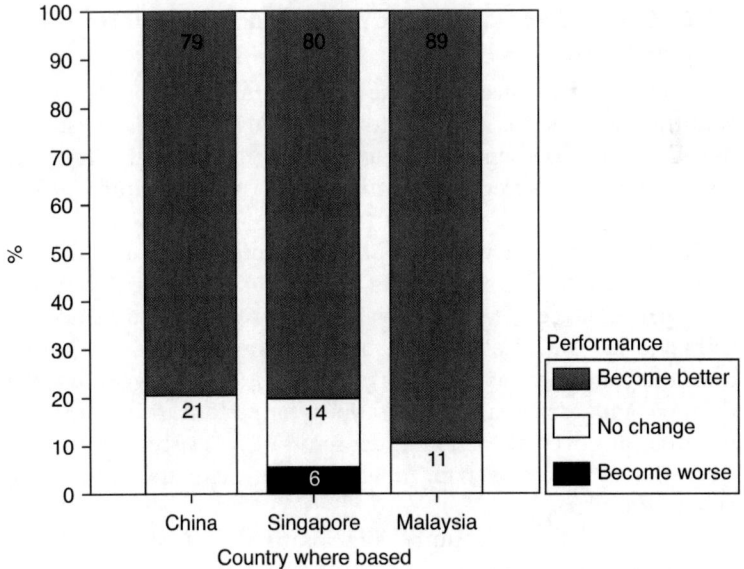

Figure 5.7   Comparison of strategic performance

in Singapore and Malaysia had made no improvements in their strategic performance.

Furthermore, the hypothetical mean values can be used to depict roughly the level of strategic performance. If the mean value is less than 3.5, hypothetically the organisation is assigned as a low (i.e. become worse) performance type; if the mean is between 3.5 and 4.5, the organisation is assigned as a medium (i.e. no change) performance type; if the mean is above 4.5, the organisation is assigned as a high (i.e. become better) performance type (see Table 5.7).

From Table 5.7, 23 out of the 29 organisations' strategic performance had become better in China, with a mean of 5.51; 28 out of 35 organisations' strategic performance had improved in Singapore, with a slightly lower mean of 5.18; and in Malaysia, 17 out of 19 organisations had improved their strategic performance, with the highest mean of 5.53. In total, only two Singapore-based organisations had performed worse, with a mean value of 3.23. Six China-based, five Singapore-based and two Malaysia-based organisations had made no changes in their strategic performance, with mean values of 4.15, 4.09 and 3.98 respectively.

5.5.2   Comparisons of strategic performance in the three selected countries

The differences in level of strategic performance across the three countries were tested by the following ANOVA tests. First, the ANOVA results are depicted using the error bar chart to show the distributions of strategic performance across the country groups (Figure 5.8).

The error bar chart was used only as a rough guide to the data. From this chart, the means plots show the different levels of strategic performance. The mean value of strategic performance for China (5.22) and Malaysia (5.36) based organisations is greater than that for Singapore-based organisations (4.90), suggesting that Chinese and Malaysian organisations outperformed Singaporean organisations over the period under scrutiny. It is also observed that all of the error bars overlap, indicating that there are no between-group differences.

In each individual country, organisations had similar trends: strategic performance had become better over the period of the five

Table 5.7 Performance classification of organisations in China, Singapore and Malaysia

Hypothetical mean levels	Levels of strategic performance	Country where based						Total
		China		Singapore		Malaysia		
		Number of firms	Mean of performance	Number of firms	Mean of performance	Number of firms	Mean of performance	
Become worse (mean less than 3.5)	Low			2	3.23			2
No change (mean between 3.5 and 4.5)	Medium	6	4.15	5	4.09	2	3.98	13
Become better (mean greater than 4.5)	High	23	5.51	28	5.18	17	5.53	68
Total		29	5.22	35	4.90	19	5.36	83

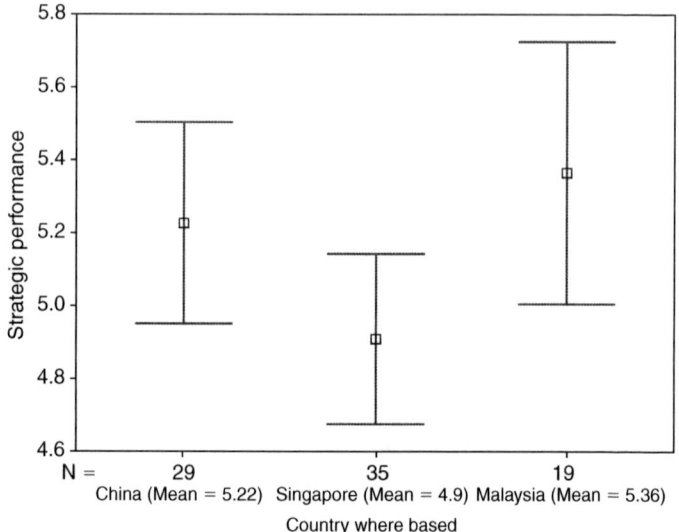

Figure 5.8   Differences of strategic performance

years examined. However, the level of improvement in strategic performance of organisations in different countries may vary and this is examined by applying the ANOVA tests.

In the ANOVA scenario, arguably, the null hypothesis that there are no differences among the groups can be rejected. This indicates that for organisations located in various countries, the levels of their strategic performance differed.

The further follow-up tests (post hoc multiple comparisons and Tukey) show that, when comparing the China group with the Singapore and Malaysia groups, there is a non-significant difference (significance is greater than 0.10). These comparisons indicated that the strategic performance for organisations situated in China did not differ from those situated in Singapore and Malaysia. When comparing the Singapore-based organisations with the Malaysia-based organisations, however, there is a marginally significant difference (significance is less than 0.10). This observation shows that the Malaysia-based organisations significantly outperformed the Singapore-based organisations over the period examined.

5.6 Summary

In each of the selected countries, organisations were classified into five strategic orientations, namely, balancer, analyser, defender, prospector and reactor. The Singapore-based organisations were more likely to employ balancer or analyser strategies than the China and Malaysia-based organisations.

Competitive strategies pursued by the participating service organisations comprise four categories, with differentiation and hybrid as the majority, low cost as the minority. Quite a few organisations fell into a non-purpose group. The similarity in the three countries is that the majority of the organisations did endeavour to pursue a differentiation-oriented strategy; organisations based in Singapore were keener on differentiations.

Four strategic positions were assigned for the participating organisations competing in the marketplace in East Asia: low-cost low-added-value; combining both price and value advantages; premium price with unique products and/or services; and high price with no added-value products and/or services. Organisations in Singapore appeared to be in a better competitive position compared with those in China and Malaysia. The majority of Singaporean organisations fell into the category of high value premium price. In Malaysia, the organisations were mainly keen on a combined competitive advantage with high value competitive price. The majority of Chinese organisations were categorised into the group of uncompetitive price and value.

For service companies operating in China, Singapore and Malaysia, there is a significantly positive relationship between strategic performance and each of the five competitive factors: quality of products or services valued by customers, reliability of technologies of products or services, safety performance of products and services, speed of response to customers, and price that customers are willing to pay.

Notably, in each country, the majority of the participating organisations had improved their strategic performance over the five-year period examined. The Malaysia-based organisations significantly outperformed the Singapore-based organisations. There was no significant difference between the China-based companies and Singapore- or Malaysia-based companies.

6
The Business Environment, Strategy and Performance

In order to review the effectiveness of a business strategy, this chapter moves on to comparisons of strategic performance when adopting different business strategies in the three selected East Asian countries. Relationships between the perceived business environment, strategic options and strategic performance are explored. The analysis can be described in terms of three aspects. First, it discovered the alignment between the perceived uncertainty and strategic options. Next, it established the relationships between strategies and the associated strategic performance. Then, correlations between the perceived environment and strategic performance were generated.

As the data are categorical or ordinal and are not normally distributed, non-parametric methods were applied in this part of the analysis. Bivariate analysis (Kerr et al., 2002) was applied to assess the relationships or conduct significance tests for the comparison of similarities and differences of various categories.

To evaluate the relationship between the two variables among the three dimensions of the perceived business environment, business strategic orientations and strategic performance, the Spearman correlation and Crosstabs with two-way contingency table analysis and chi-square tests were employed. Kruskal–Wallis tests were also applied when testing the differences of the perceived environmental uncertainty and strategic performance by strategic groups. A number of box plot graphs and scatter diagrams were also applied to highlight the pattern of differences and correlations.

6.1   Business environment and strategy alignment

Following Chapters 4 and 5, the remainder of the ten propositions are examined in this chapter. According to Proposition 8 presented in Chapter 1, for oil and gas service organisations operating in East Asia, managerial perceptions of environmental uncertainty will vary in association with the types of business strategies.

In order to examine whether the perceived environmental uncertainty differed for the participating service organisations with different strategic options, a box plot graph was used to explore the results informatively (Figure 6.1(a)). Balancers appeared to have a relatively lower degree of perceived environmental uncertainty than organisations in other strategic categories as they had the lowest medians. Reactors and prospectors tended to have a higher level of perceived

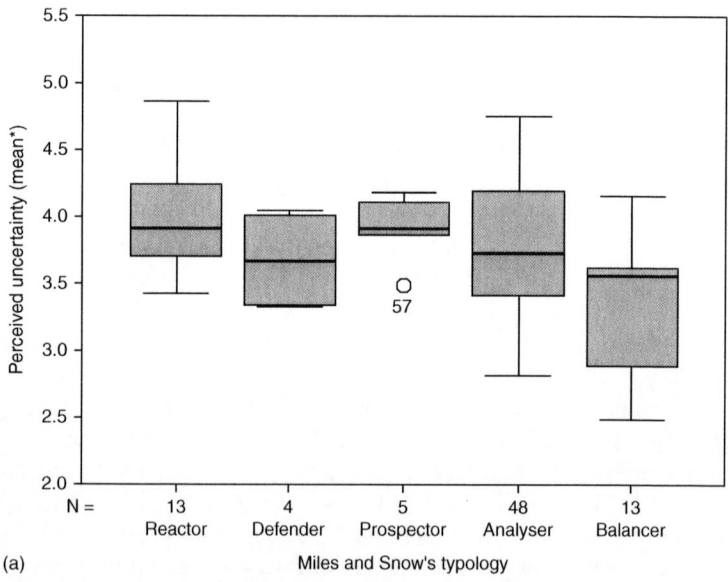

(a)

Figure 6.1(a)   Perceived uncertainty by business strategies

* In order to obtain a picture to better distinguish different levels of the perceived uncertainty by Miles and Snow's typology, the mean value of the assessment of environmental factors is adopted for assigning the perceived uncertainty.

uncertainty. The managerial perception of defenders and analysers on environmental uncertainty was relatively lower than that of reactors and prospectors.

The Kruskal–Wallis test (Figure 6.1(b)) indicates that there is a marginally significant difference in the medians, X^2 (4, $N = 83$) = 8.308, $p = 0.081$. This means that the hypothesis that all the medians of perceived environmental uncertainty among different strategic groups are equal can be rejected at the 10 per cent significant level. Because the overall test is significant, pairwise comparisons among the five groups were conducted.

In the following analysis, five generic strategies were assigned into five groups: Group 1 – reactors, Group 2 – defenders, Group 3 – prospectors, Group 4 – analysers and Group 5 – balancers.

Ranks

	Miles and Snow typology	N	Mean rank
Perceived uncertainty (mean)	Reactor	13	52.65
	Defender	4	35.50
	Prospector	5	50.70
	Analyser	48	42.71
	Balancer	13	27.38
	Total	83	

Test statistics [a,b]

	Perceived uncertainty (mean)
Chi-square	8.308
df	4
Asymptotic significance	0.081

a. Kruskal–Wallis test.
b. Grouping variable: Miles and Snow typology.

Figure 6.1(b)   Kruskal–Wallis test ($N = 83$)*

* Organisations which were less than five years old were omitted as it was decided to evaluate strategic performance over a period of at least five years. In order to avoid a biased assessment, responses of the functional managers and from other countries were also omitted. As a result, in total 83 organisations were selected to run this test.

Ranks

Miles and Snow typology		N	Mean rank	Sum of ranks
Perceived uncertainty (mean)	Reactor	13	17.27	224.50
	Balancer	13	9.73	126.50
	Total	26		

Test statistics[b]

	Perceived uncertainty (mean)
Mann–Whitney U	35.500
Wilcoxon W	126.500
Z	−2.514
Asymptotic significance (2-tailed)	0.012
Exact significance [2*(1-tailed significance)]	0.010[a]

a. Not corrected for ties.
b. Grouping variable: Miles and Snow typology.

Ranks

Miles and Snow typology		N	Mean rank	Sum of ranks
Perceived uncertainty (mean)	Prospector	5	13.10	65.50
	Balancer	13	8.12	105.50
	Total	18		

Test statistics[b]

	Perceived uncertainty (mean)
Mann–Whitney U	14.500
Wilcoxon W	105.500
Z	−1.776
Asymptotic significance (2-tailed)	0.076
Exact significance [2*(1-tailed Sig.)]	0.075[a]

a. Not corrected for ties.
b. Grouping variable: Miles and Snow typology.

Figure 6.1(c) Mann–Whitney U-test

Ranks

	Miles and Snow typology	N	Mean rank	Sum of ranks
Perceived uncertainty (mean)	Analyser	48	33.45	1605.50
	Balancer	13	21.96	285.50
	Total	61		

Test statistics[a]

	Perceived uncertainty (mean)
Mann–Whitney U	194.500
Wilcoxon W	285.500
Z	–2.070
Asymptotic significance (2-tailed)	0.038

a. Grouping variable: Miles and Snow typology.

Figure 6.1(c) (Continued)

Using the Mann–Whitney U-tests (Figure 6.1(c)), three pairs of comparisons between two of the five groups are significantly different: Group 5 and 1, 3, 4 respectively. No significant results were obtained to prove significant differences between Group 2, the defenders and Group 5, the balancers.

As a result, balancers had a significantly different perception of environmental uncertainty from the organisations in other categories of business strategies. Compared with reactors, prospectors and analysers, the degree of environmental uncertainty perceived by balancers was lower. However, balancers and defenders shared similar views on the degree of perceived uncertainty.

The outcomes indicate that balancers felt relatively more comfortable with the business environment than prospectors, reactors and analysers. Reactors appeared relatively too 'lazy' to monitor what was going on in the business environment where they operated.

The results do not provide sufficient evidence to reject the hypothesis that the medians of perceived environmental uncertainty among reactor, defender, prospector and analyser strategic groups are equal. This suggests that, for organisations having any of these four types of strategies, their managerial perceptions on environmental uncertainty may not be dissimilar.

In order to investigate the assumptions regarding the association between the strategic options and perceived environmental uncertainty, chi-square tests with Crosstabs were applied. Where appropriate, to test the strength of association, Cramer's values were used as each of the two variables has more than two categories.

The following chi-square test (Table 6.1) was computed to determine whether there is an association between two categorical variables (i.e. generic strategic directions and perceived environmental uncertainty).

In this study, five categories were designed initially for assigning business strategies. In order to obtain a more expected count (ideally greater than 5) in the cross-tabulation table and reliable chi-square test results, the categories of business strategies were recoded into two combined categories: using multiple business strategies or not using multiple business strategies.

As shown in Table 6.1, in order to boost the proportion of cases falling into each variable, the combined categories of the first variable are non-multiple business strategy users (i.e. a combination of prospectors, defenders and reactors) and multiple business strategy users (i.e. a combination of balancers and analysers); the combined categories of the second variable are uncertainty (i.e. a mix of tending to be uncertain, uncertain and very uncertain) and not uncertainty (i.e. a mix of neutral, tending to be certain, certain and very certain).

From the analytical results shown in Table 6.1, the significance value is 0.247. Consequently, the hypothesis that the environment–strategy variables are independent is not rejected and the hypothesis that they are in some way related cannot be accepted. Hence, generic business strategies do not provide any hints as to whether or not the perceived business environment could be uncertain.

As such, no evidence can be provided to support the assumption that the perceived uncertainty varies when applying different strategic types; or in other words, none of the generic business

Table 6.1 Strategies–environment correlation test

			Types of business strategy		Total
			Non-multiple business strategy	Multiple business strategy	
Perceived uncertainty	Uncertainty	Count	13	31	44
		Expected count	10.6	33.4	44.0
		% within perceived uncertainty	29.5	70.5	100.0
		% within Miles and Snow	59.1	44.9	48.4
		% of Total	14.3	34.1	48.4
	Not uncertainty	Count	9	38	47
		Expected count	11.4	35.6	47.0
		% within perceived uncertainty	19.1	80.9	100.0
		% within Miles and Snow	40.9	55.1	51.6
		% of Total	9.9	41.8	51.6
Total		Count	22	69	91
		Expected count	22.0	69.0	91.0
		% within perceived uncertainty	24.2	75.8	100.0
		% within Miles and Snow	100.0	100.0	100.0
		% of Total	24.2	75.8	100.0

(*Continued*)

Table 6.1   Continued

Chi-square tests	Value	df	Asymptotic significance (2-sided)	Exact significance (2-sided)	Exact significance (1-sided)
Pearson chi-square	1.340[a]	1	0.247		
Continuity correction[b]	0.833	1	0.361		
Likelihood ratio	1.344	1	0.246		
Fisher's exact test				0.328	0.181
Linear-by-linear association	1.325	1	0.250		
N of valid cases	91				

[a] 0 cells (0%) have an expected count less than 5. The minimum expected count is 10.64.
[b] Computed only for a 2 × 2 table.

strategies or competitive positions was found to be significantly related to the perceived environmental uncertainty. Consequently, the results fail to prove the contention of Proposition 8 that there is a relationship between perceived business environmental uncertainty and types of business strategic options. Hence, no matter what generic strategic directions a service organisation follows, the level of perceived environmental uncertainty may not be associated with the strategic options. Alternatively, managerial perceptions on the business environment may not have a significant impact on the selection of oil service companies' long-term business strategies.

6.2   Environment and strategic performance relationship

This section is concerned with the examination of Proposition 9 that there will be relationships between the perceived business environment and strategic performance for oil and gas service organisations operating in East Asia.

In order to evaluate whether the perceived environmental uncertainty, complexity, dynamism and hostility are associated with strategic performance, Spearman's correlations were processed. Since the two variables were ordinal data, the Spearman rank-order correlation coefficient was computed.

The results were presented by a scatter plot graph shown in Figure 6.2. It displays the result that there is a negative relationship between strategic performance and perceived environmental hostility and uncertainty (slope down), yet there is no relationship between strategic performance and perceived environmental dynamism and complexity.

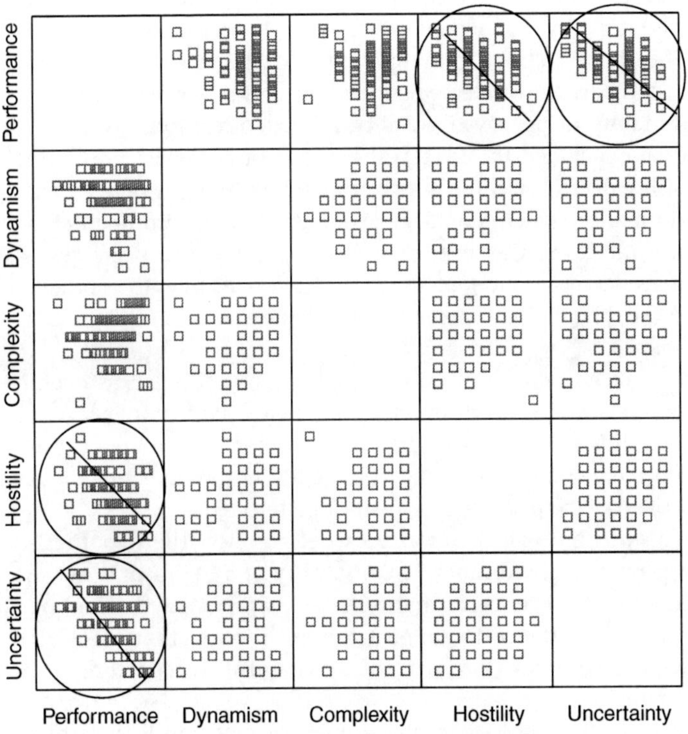

Figure 6.2 Perceived environment and strategic performance

It was predicted that strategy performance would correlate significantly with perceived environmental uncertainty, complexity, dynamism and hostility. Therefore, the test for these variables should be two-tailed. SPSS output provides a matrix of the correlation coefficients for the five variables (Table 6.2). There is a significant negative correlation between strategic performance and perceived environmental uncertainty ($r = -0.412$, $p < 0.001$). Strategic performance is also negatively related to perceived hostility ($r = -0.324$, $p = 0.001$). Hence, it can be confidently stated that the relationship between strategic performance and perceived uncertainty or hostility is legitimate.

Since the significance value for their association coefficient is less than 0.01, the results provide sufficient evidence to reject the hypothesis that the perceived environmental uncertainty or the perceived hostility is independent of strategic performance.

The above research findings show that there is a significant relationship between strategic performance and perceived environmental uncertainty or perceived hostility. As the correlations are negative, it can be concluded that, when the degree of perceived uncertainty or hostility decreases, there is a corresponding improvement in strategic performance; when the degree of perceived environmental uncertainty increases, the level of strategic performance decreases; when managerial perception of the business environment becomes more pleasant, there is also a coordinated better outcome in strategic performance.

Conversely, the significance value for the correlation coefficient between strategic performance and perceived complexity or perceived dynamism is more than 0.10. Therefore, the hypothesis that the perceived environmental complexity or dynamism is independent of strategic performance cannot be rejected.

No sufficient evidence was provided to reject the hypothesis that the perceived environmental complexity or dynamism is not associated with strategic performance ($p > 0.05$). Hence, the findings fail to provide evidence to support the assumption that the degree of perceived complexity or perceived dynamism is related to the level of strategic performance.

Overall, the results partially support Proposition 9 that there is a relationship between the perceived business environmental dimensions and strategic performance for oil and gas service organisations

Table 6.2 Environment–strategic performance relationships

		Correlations				
		Perceived uncertainty	Perceived hostility	Perceived complexity	Perceived dynamism	Strategic performance
Spearman's rho	Perceived uncertainty	Correlation coefficient	1.000	0.269**	-0.144	0.209*	-0.412**
		Significance (2-tailed)	–	0.007	0.157	0.039	0.000
		N	98	98	98	98	98
	Perceived hostility	Correlation coefficient	0.269**	1.000	0.064	0.234*	-0.324**
		Significance (2-tailed)	0.007	–	0.530	0.020	0.001
		N	98	98	98	98	98
	Perceived complexity	Correlation coefficient	-0.144	0.064	1.000	0.236*	0.137
		Significance (2-tailed)	0.157	0.530	–	0.019	0.179
		N	98	98	98	98	98
	Perceived dynamism	Correlation coefficient	0.209*	0.234*	0.236*	1.000	-0.042
		Significance (2-tailed)	0.039	0.020	0.019	–	0.684
		N	98	98	98	98	98
	Strategic performance	Correlation coefficient	-0.412**	-0.324**	0.137	-0.042	1.000
		Significance (2-tailed)	0.000	0.001	0.179	0.684	–
		N	98	98	98	98	98

* Correlation is significant at the 0.05 level (2-tailed).
** Correlation is significant at the 0.01 level (2-tailed).

operating in East Asia. Strategic performance is negatively related to perceived environmental uncertainty or hostility. In this sense, the higher the perceived environmental uncertainty or hostility, the weaker the strategic performance.

6.3   Comparisons of strategic performance by strategic groups

In this section, attention is paid to examine Proposition 10 that, for oil and gas service organisations operating in East Asian countries like China, Singapore and Malaysia, strategic performance will be associated with their strategic business orientations.

6.3.1   Strategic performance with business strategies

It was assumed that, for oil and gas service organisations pursuing different strategies, their strategic performance would be different. In order to examine whether strategic performance differed for the participating service organisations with different strategic options, Kruskal–Wallis tests were applied. Where appropriate, follow-up pairwise Mann–Whitney U-tests were applied if the overall test was significant. Before conducting the significance test, a box plot graph was used to depict differences (Figure 6.3).

The strategic performance of analysers and balancers appeared to be generally higher than the organisations in other strategic categories as they had higher medians and quartiles. It is observed that having a reactor strategy did not give the organisations a relatively poor strategic performance. Rather, organisations with defender and prospector strategies appeared to perform poorly.

The Kruskal–Wallis test (Table 6.3) indicates that there is a marginally significant difference in the medians, $X^2 = 8.712$ (df = 4), $p = 0.059$. This means that the hypothesis that all the medians of strategic performance among different strategic groups are equal can be rejected at the 10 per cent significant level.

Because the overall test is significant, pairwise comparisons among the five groups were conducted. Using the Mann–Whitney U-tests, four pairs of comparisons between the two of the five groups are significantly different: Groups 1 and 4, and Groups 1 and 5, Groups 2 and 4, Groups 2 and 5 (Table 6.4).

The evidence proves that analysers or balancers outperformed defenders and reactors in the East Asian market. This suggests

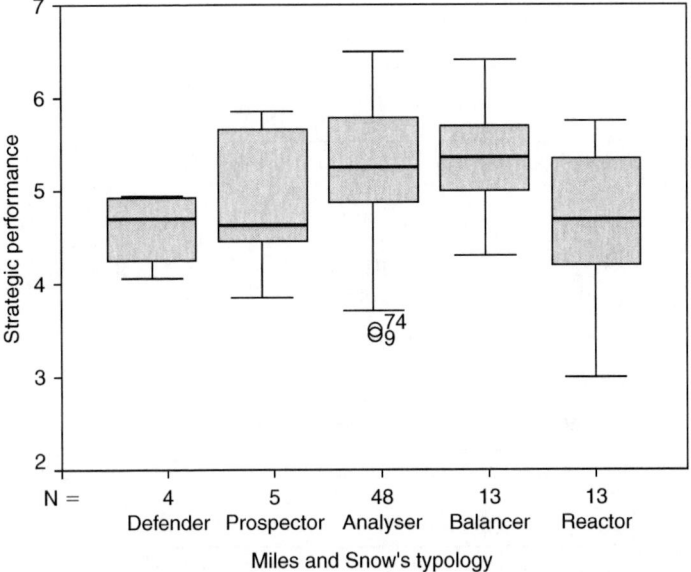

Figure 6.3   Strategic performance by business strategies

that using either an analyser or balancer strategy can guarantee organisations better success than those using a defender or reactor strategy.

The results do not provide sufficient evidence to reject the hypothesis that the medians of strategic performance among defender, prospector and reactor strategic groups are equal. This suggests that there are no differences of strategic performance if an organisation pursues a defender, prospector or reactor strategy. What is more, no significant results were obtained to prove the differences between prospectors and analysers or between prospectors and balancers. Hence, as compared to analysers or balancers, having a prospector strategy does not yield an organisation a significantly different strategic performance.

In short, the results partially support the assumption that balancers and analysers outperform defenders or reactors. The analysis fails to prove that reactors are associated with a relatively poor strategic performance, while the strategic performance of defenders

Table 6.3   Strategic performance by business strategies

Ranks

	Miles and Snow typology	N	Mean rank
Strategic performance	Reactor	13	30.92
	Defender	4	19.75
	Prospector	5	34.50
	Analyser	48	45.44
	Balancer	13	50.12
	Total	83	

Test statistics[a,b]

	Strategic performance
Chi-square	9.070
df	4
Asymptotic significance	0.059

[a] Kruskal–Wallis test.
[b] Grouping variable: Miles and Snow typology.

and prospectors stands in the middle among the five categories of strategic orientations.

6.3.2   Strategic performance with competitive strategies

A box plot graph (Figure 6.4(a)) shows that among the four competitive strategies, hybrid organisations had a relatively higher strategic performance whereas low-cost organisations performed relatively poorly. Differentiation organisations had a sound performance and no-purpose organisations did not perform the worst even though they did not adopt any competitive strategies. Low-cost organisations appeared to perform more poorly than no-purpose organisations.

It was assumed that for the sample organisations that adopted different competitive strategies, the strategic performance would not be the same. In order to examine whether strategic performance is different with various competitive strategies, the Kruskal–Wallis tests were applied. The results are shown in Figure 6.4(b).

Table 6.4 Mann–Whitney *U*-tests for business strategies and strategic performance

		Ranks				Test statistics[a]	
		Miles and Snow typology	N	Mean rank	Sum of ranks		Strategic performance
I *Groups* *1 and 4*	Strategic performance	Reactor	13	22.69	295.00	Mann–Whitney *U*	204.000
		Analyser	48	33.25	1596.00	Wilcoxon *W*	295.000
		Total	61			*Z*	−1.903
						Asymptotic significance (2-tailed)	0.057

[a]Grouping variable: Miles and Snow typology.

		Ranks				Test statistics[b]	
		Miles and Snow typology	N	Mean rank	Sum of ranks		Strategic performance
II *Groups* *1 and 5*	Strategic performance	Reactor	13	10.46	136.00	Mann–Whitney *U*	45.000
		Balancer	13	16.54	215.00	Wilcoxon *W*	136.000
		Total	26			*Z*	−2.027
						Asymptotic significance (2-tailed)	0.043
						Exact significance [2*(1-tailed significance)]	0.044[a]

[a]Not corrected for ties.
[b]Grouping variable: Miles and Snow typology.

(Continued)

Table 6.4 Continued

III
Groups 2 and 4

| | Ranks | | | | Test statistics[b] |
	Miles and Snow typology	N	Mean rank	Sum of ranks	Strategic performance
Strategic performance	Defender	4	11.38	45.50	
	Analyser	48	27.76	1332.50	
	Total	52			
Mann–Whitney U					35.500
Wilcoxon W					45.500
Z					−2.079
Asymptotic significance (2-tailed)					0.038
Exact significance [2*(1-tailed significance)]					0.034[a]

[a] Not corrected for ties.
[b] Grouping variable: Miles and Snow assignment.

IV
Groups 2 and 5

| | Ranks | | | | Test statistics[b] |
	Miles and Snow typology	N	Mean rank	Sum of ranks	Strategic performance
Strategic performance	Defender	4	3.63	14.50	
	Balancer	13	10.65	138.50	
	Total	17			
Mann–Whitney U					4.500
Wilcoxon W					14.500
Z					−2.439
Asymptotic significance (2-tailed)					0.015
Exact significance [2*(1-tailed significance)]					0.010[a]

[a] Not corrected for ties.
[b] Grouping variable: Miles and Snow assignment.

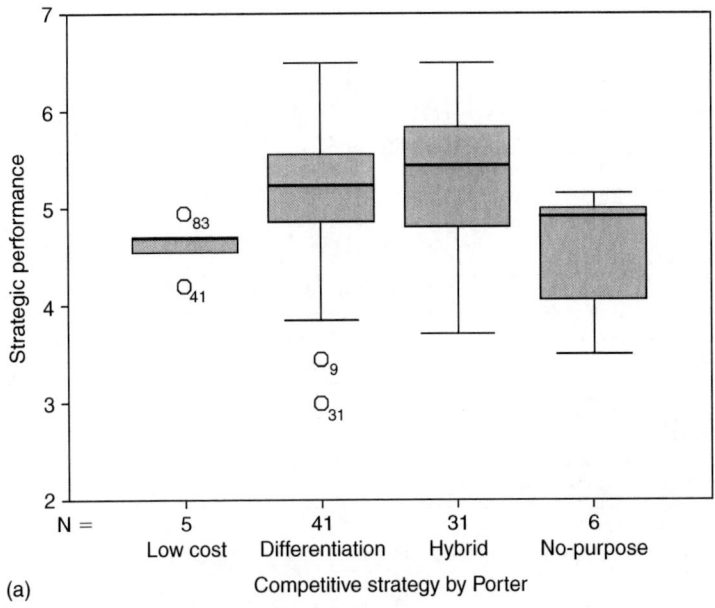

Figure 6.4(a) Strategic performance by competitive strategies

Ranks

	Competitive	N	Mean rank
Strategic performance	Low cost	5	20.60
	Differentiation	41	42.07
	Hybrid	31	48.82
	No-purpose	6	24.08
	Total	83	

Test statistics[a,b]

	Strategic performance
Chi-square	9.750
df	3
Asymptotic significance	0.021

a. Kruskal–Wallis test.
b. Grouping variable: competitive strategy by Porter.

Figure 6.4(b) Kruskal–Wallis test

Ranks

	Competitive	N	Mean rank	Sum of ranks
Strategic performance	Low cost	5	11.60	58.00
	Differentiation	41	24.95	1023.00
	Total	46		

Test statistics[b]

	Strategic performance
Mann–Whitney U	43.000
Wilcoxon W	58.000
Z	−2.102
Asymptotic significance (2-tailed)	0.036
Exact significance [2*(1-tailed significance)]	0.034[a]

a. Not corrected for ties.
b. Grouping variable: competitive strategy by Porter.

Ranks

	Competitive	N	Mean rank	Sum of ranks
Strategic performance	Low cost	5	9.60	48.00
	Hybrid	31	19.94	618.00
	Total	36		

Test statistics[b]

	Strategic performance
Mann–Whitney U	33.000
Wilcoxon W	48.000
Z	−2.038
Asymptotic significance (2-tailed)	0.042
Exact significance [2*(1-tailed significance)]	0.041[a]

a. Not corrected for ties.
b. Grouping variable: competitive strategy by Porter.

Figure 6.4(c) Mann–Whitney U-test

Ranks

	Competitive	N	Mean rank	Sum of ranks
Strategic performance	Hybrid	31	20.63	639.50
	No-purpose	6	10.58	63.50
	Total	37		

Test statistics[b]

	Strategic performance
Mann–Whitney U	42.500
Wilcoxon W	63.500
Z	−2.082
Asymptotic significance (2-tailed)	0.037
Exact significance [2*(1-tailed significance)]	0.035[a]

a. Not corrected for ties.
b. Grouping variable: competitive strategy by Porter.

Figure 6.4(c) (Continued)

The Kruskal–Wallis test indicates that there is a significant difference in the medians, X^2 = 9.750 (df = 3), p = 0.021. Hence, the hypothesis that the strategic performance differs for organisations with different competitive strategies cannot be rejected.

Since the overall test is significant, using the Mann–Whitney U- test, pairwise comparisons among the four groups were conducted. Comparisons between two of the four groups were made. Groups 1, 2, 3 and 4 are low cost, differentiation, hybrid and no-purpose respectively. The comparisons between low-cost and differentiation groups, low-cost and hybrid groups, and hybrid and no-purpose groups are significant (Figure 6.4(c)).

The significant results prove that differentiation or hybrid organisations outperformed low-cost organisations. This finding suggests that the low-cost strategy is not an ideal option for oil and gas service organisations. Meanwhile, hybrid organisations also outperformed those without any competitive strategies.

There is no significant evidence to prove that strategic performance differs between no-purpose and low-cost or differentiation organisations. This indicates that those organisations not pursuing any competitive strategies may have a similar performance achievement as those pursuing a low-cost or differentiation strategy. Differentiation organisations also did not perform significantly differently from hybrid organisations.

The results do not support the assumption that having a competitive strategy yields the organisation a higher level of strategic performance, while pursuing none of the three generic competitive strategies produces a relatively poor strategic performance. The assumption that hybrid organisations yield a higher level of strategic performance than low-cost and non-purpose strategic groups was partially proved.

6.3.3   Strategic performance with strategic positions

It was also assumed that strategic performance varied when organisations had been in different strategic positions in the marketplace. To test the differences of strategic performance, the same tests as above were conducted. A box plot diagram was used for preliminary exploration of the results (Figure 6.5(a)).

From the box plot graph, it is observed that organisations with competitive price and low or moderate value produced the highest performance, while uncompetitive value and price organisations were in the category of the lowest strategic performance. Strategic performance for organisations with high value and competitive or premium price stood in the middle.

The overall Kruskal–Wallis test indicates that there is a significant difference in the medians, $X^2 = 11.722$ (df = 3), $p = 0.008$ (Figure 6.5(b)). Hence, strategic performance for organisations in different strategic positions varied.

Pairwise comparisons by the Mann–Whitney U-test between two of the four groups were also made (Figure 6.5(c)). Three pairs of comparisons are significant. Low or moderate value with competitive (low and moderate) price organisations, organisations with high value competitive price, and organisations with high value premium price outperformed those with uncompetitive value and price. Obviously, being in a strategic position of providing low customer added value with a high price will definitely bring organisations a poor strategic performance.

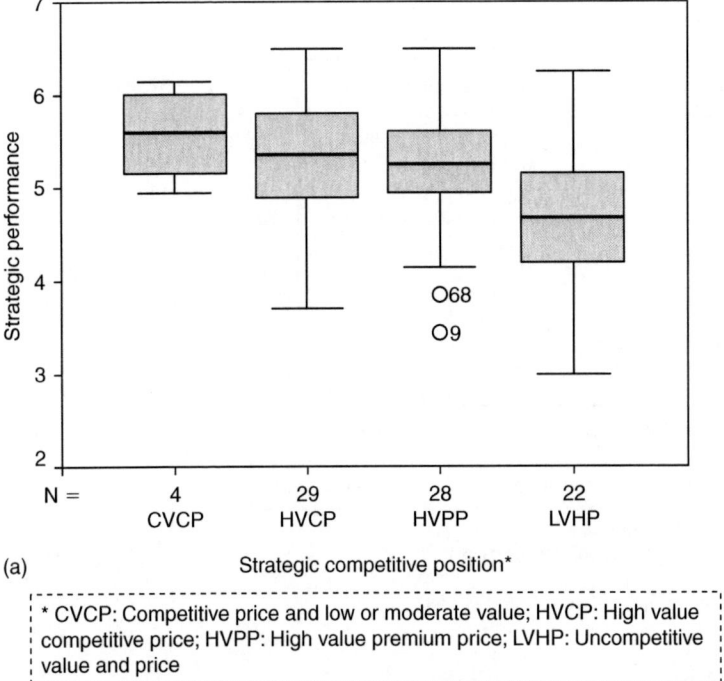

(a)   Strategic competitive position*

> * CVCP: Competitive price and low or moderate value; HVCP: High value
> competitive price; HVPP: High value premium price; LVHP: Uncompetitive
> value and price

Figure 6.5(a) Strategic performance by strategic positions

There is no significant difference in strategic performance amongst the other three categories of strategic positions apart from the uncompetitive value and price situation.

Meanwhile, the initial assumption that hybrid organisations outperform differentiation organisations was not proved. The assumption that firms with a differentiation strategy outperform those with a low-cost strategy was supported. This finding suggests that a differentiation strategy is a more appropriate option than a low-cost one for service organisations operating in East Asia.

The finding supports the assumption that organisations falling into the category of low value high price would have a poorer strategic performance than other organisations. The results do not support the assumption that organisations with a high value competitive

Ranks

Strategy clock		N	Mean rank
Strategic performance	Low (or moderate) value competitive price	4	57.88
	High value competitive (low and moderate price)	29	48.29
	High value premium price	28	44.43
	Uncompetitive value and price	22	27.73
	Total	83	

Test statistics[a,b]

	Strategic performance
Chi-square	11.722
df	3
Asymptotic significance	0.008

a. Kruskal–Wallis test.
b. Grouping variable: strategy clock.

Figure 6.5(b)   Kruskal–Wallis test

Ranks

Strategy clock		N	Mean rank	Sum of ranks
Strategic performance	High value premium price	28	30.05	841.50
	Uncompetitive value and price	22	19.70	433.50
	Total	50		

Test statistics[a]

	Strategic performance
Mann–Whitney U	180.500
Wilcoxon W	433.500
Z	−2.494
Asymptotic significance (2-tailed)	0.013

a. Grouping variable: strategy clock.

Figure 6.5(c)   Mann–Whitney U-test

Ranks

	Strategy clock positions	N	Mean rank	Sum of ranks
Strategic performance	Low (or moderate) value competitive price	4	20.75	83.00
	Uncompetitive value and price	22	12.18	268.00
	Total	26		

Test statistics[b]

	Strategic performance
Mann–Whitney U	15.000
Wilcoxon W	268.000
Z	−2.064
Asymptotic significance (2-tailed)	0.039
Exact significance [2*(1-tailed significance)]	0.039[a]

a. Not corrected for ties.
b. Grouping variable: strategy clock positions.

Ranks

	Strategy clock	N	Mean rank	Sum of ranks
Strategic performance	High value competitive (low and moderate) price	29	31.43	911.50
	Uncompetitive value and price	22	18.84	414.50
	Total	51		

Test statistics[a]

	Strategic performance
Mann–Whitney U	161.500
Wilcoxon W	414.500
Z	−2.999
Asymptotic significance (2-tailed)	0.003

a. Grouping variable: strategy clock.

Figure 6.5(c) (Continued)

(low and moderate) price position would outperform companies in the other three categories of competitive positions.

Overall, the statistics show that Proposition 10 supported the point that a good or bad strategic performance is associated with certain strategic orientations adopted for service businesses in East Asia. To conclude:

• better strategic performance is associated with balancer or analyser strategies rather than with other types of strategy;
• better strategic performance is associated with differentiation-oriented strategies rather than with other types of competitive strategy;
• weaker strategic performance is associated with low customer added value premium prices rather than with other types of strategic competition position.

6.4   Cross-national correlation comparisons

In earlier sections, the relationships were set up between the two variables amongst strategies, and performance and the perceived business environment. This section intends to compare the differences of those established associations in various country groups.

6.4.1   Cross-national strategy–performance association

The relationship of two variables – business strategies and strategic performance – is examined in this section. A cross-tabulation table was applied and it contains the number of cases that falls into each combination of categories. Data were split into the China, Singapore and Malaysia groups (Table 6.5).

Initially, there were five categories for business strategies and seven ordinal categories for strategic performance. In order to obtain a more reliable outcome (ideally expected count greater than 5) in the cross-tabulation table, the categories of business strategies and strategic performance were recoded.

For business strategies, there were two combined categories of whether or not multiple business strategies were used: yes (i.e. they were either balancers or analysers) or no (i.e. they were defenders, prospectors or reactors). Three combined categories for the level

Table 6.5 Comparison of performance–strategy correlation

Country where based			Strategic performance			Total
			Become worse	No change	Become better	
China	Multiple strategies?	Yes	Count		3	18	21
			Expected count		4.3	16.7	21.0
			% within multiple strategies?		14.3	85.7	100.0
			% within strategic performance		50.0	78.3	72.4
			% of Total		10.3	62.1	72.4
		No	Count		3	5	8
			Expected count		1.7	6.3	8.0
			% within multiple strategies?		37.5	62.5	100.0
			% within strategic performance		50.0	21.7	27.6
			% of Total		10.3	17.2	27.6
	Total		Count		6	23	29
			Expected count		6.0	23.0	29.0
			% within multiple strategies?		20.7	79.3	100.0
			% within strategic performance		100.0	100.0	100.0
			% of Total		20.7	79.3	100.0

(Continued)

Table 6.5 Continued

Country where based	Multiple strategies?		Strategic performance			Total
			Become worse	No change	Become better	
Singapore	Yes	Count	1	2	25	28
		Expected count	1.6	4.0	22.4	28.0
		% within multiple strategies?	3.6	7.1	89.3	100.0
		% within strategic performance	50.0	40.0	89.3	80.0
		% of Total	2.9	5.7	71.4	80.0
	No	Count	1	3	3	7
		Expected count	0.4	1.0	5.6	7.0
		% within multiple strategies?	14.3	42.9	42.9	100.0
		% within strategic performance	50.0	60.0	10.7	20.0
		% of Total	2.9	8.6	8.6	20.0
	Total	Count	2	5	28	35
		Expected count	2.0	5.0	28.0	35.0
		% within multiple strategies?	5.7	14.3	80.0	100.0
		% within strategic performance	100.0	100.0	100.0	100.0
		% of Total	5.7	14.3	80.0	100.0

Malaysia						
Multiple strategies?	Yes	Count	1	11	12	
		Expected count	1.3	10.7	12.0	
		% within multiple strategies?	8.3	91.7	100.0	
		% within strategic performance	50.0	64.7	63.2	
		% of Total	5.3	57.9	63.2	
	No	Count	1	6	7	
		Expected count	0.7	6.3	7.0	
		% within multiple strategies?	14.3	85.7	100.0	
		% within strategic performance	50.0	35.3	36.8	
		% of Total	5.3	31.6	36.8	
Total		Count	2	17	19	
		Expected count	2.0	17.0	19.0	
		% within multiple strategies?	10.5	89.5	100.0	
		% within strategic performance	100.0	100.0	100.0	
		% of Total	10.5	89.5	100.0	

of strategic performance were formed as: become better (including tend to be better, better and much better categories), no change, and become worse (including tend to be worse, worse, much worse categories).

Over the five-year period (1997–2001) under scrutiny, a total of 21 China-based organisations used multiple business strategies (72.4 per cent of the total). Of those, 18 organisations' strategic performance had become better (85.7 per cent of the total that used multiple strategies) and only three organisations' strategic performance remained the same (14.3 per cent of the total that used multiple strategies). Eight organisations did not use any type of multiple business strategies such as balancer or analyser (27.6 per cent of the total). Of those, five organisations' strategic performance had become better (62.5 per cent of the total that did not use multiple strategies) and only three organisations' strategic performance remained the same (37.5 per cent of the total that did not use multiple strategies).

In summary, more service organisations in China employed multiple business strategies (72.4 per cent) than those that did not (27.6 per cent). However, regardless of selection of the use of multiple businesses strategies, most of the China-based organisations had improved performance (79.3 per cent of the total).

For the Malaysia group, the data can be summarised in a similar way and the results observed are similar to the China group. A total of 12 Malaysia-based organisations used multiple business strategies (63.2 per cent of the total). Of those, 11 organisations' strategic performance had become better (91.7 per cent of the total that used multiple strategies) and only one organisation's strategic performance remained the same (8.3 per cent of the total that used multiple strategies).

Seven organisations did not use any type of multiple business strategies (36.8 per cent of the total). Of those, six organisations' strategic performance had become better (85.7 per cent of the total that did not use multiple strategies) and only one organisation's strategic performance remained the same (14.3 per cent of the total that did not use multiple strategies) over the five-year period.

Hence, more Malaysia-based organisations (63.2 per cent) employed multiple business strategies (i.e. either balancers or analysers) than those that did not (i.e. defenders, prospectors or reactors) (36.8 per cent). In both categories, a large majority had improved their strategic performance.

In short, organisations operating in Malaysia appeared to have improved strategic performance over the five years examined no matter whether or not they chose to use multiple business strategies (89.5 per cent of the total).

With regard to the empirical results from Singapore, an extra category (i.e. become worse) on the level of strategic performance appears in the summary data for this country group (see Table 6.5).

For Singapore-based organisations, 28 used multiple business strategies (80 per cent of the total). Of those, 25 organisations' strategic performance had become better (89.3 per cent), two organisations' strategic performance had remained the same (7.1 per cent) and one indicated that its strategic performance had become worse (3.6 per cent).

Seven Singapore-based organisations did not use any type of multiple business strategies (20 per cent of the total). Of those, three organisations' strategic performance had become better (42.9 per cent), three organisations' strategic performance had remained the same (42.9 per cent) and one had performed worse (14.3 per cent) over the five-year period.

To summarise, when multiple business strategies were employed, most Singapore organisations performed better, but when multiple business strategies were not introduced, most Singapore organisations' performance either remained the same or became worse. However, regardless of the level of strategic performance, a large majority of organisations used multiple business strategies.

In order to examine the significance of the strategy–performance association in the selected countries, further statistic tests were conducted. The chi-square test examines whether there is an association between two categorical variables: the level of strategic performance and multiple-strategy users. The Pearson chi-square statistic tests whether the two variables are independent (Table 6.6).

For both the China and Malaysia groups, the chi-square statistic is observed to be insignificant ($p > 0.05$), indicating that the level of strategic performance is not affected by whether multiple strategies were used. This finding reflects the fact that the level of strategic performance for the China and Malaysia groups was identical in the two strategic options.

In contrast, for the Singapore group, the chi-square statistic observed is significant ($p < 0.05$), and the hypothesis that the variables are independent could be rejected. This means that in Singapore

Table 6.6　Significance of strategy–performance association[a,b]

Country where based		Value	Df	Asymptotic significance (2-sided)
China[c]	Pearson chi-square	1.903	1	0.168
	N of valid cases	29		
Singapore[d]	Pearson chi-square	7.634	2	0.022
	N of valid cases	35		
Malaysia[e]	Pearson chi-square	0.166	1	0.683
	N of valid cases	19		

[a] Computed only for a 2×2 table.
[b] As more than 20 per cent of expected frequencies below 5, the result may be a loss of statistical power. Proportionately small differences in cell frequencies could result in statistically significant associations between variables (Field, 2000). For these reasons, the observed results are only used as exploratory findings for the study.
[c] 2 cells (50.0%) have expected count less than 5. The minimum expected count is 1.66.
[d] 4 cells (66.7%) have expected count less than 5. The minimum expected count is 0.40.
[e] 2 cells (50.0%) have expected count less than 5. The minimum expected count is 0.74.

strategic performance was in some way associated with the choice of business strategies. Better strategic performance is significantly associated with multiple business strategies, whereas poor strategic performance is associated with those users of non-multiple business strategies.

In Figure 6.6, the outcome of Cramer's statistics highlights these differences: the proportion of levels of strategic performance in the two strategic conditions is opposite for the Singapore group, but identical in the China and Malaysia groups.

For the Singapore group, Cramer's statistic is 0.467 out of a possible maximum value of 1. This represents a relatively strong association between the level of strategic performance and whether the organisations used multiple strategies or not. This value is significant ($p < 0.05$), indicating that strength of the relationship is significant.

For the China and Malaysia groups, Cramer's statistic is 0.256 and 0.094 respectively, indicating the association between the two variables is weak for the China and Malaysia organisations. The probability that this value is a chance result is 0.168 for China and 0.683 for Malaysia.

These results confirm what the chi-square test shows: the level of strategic performance is significantly related to whether or not

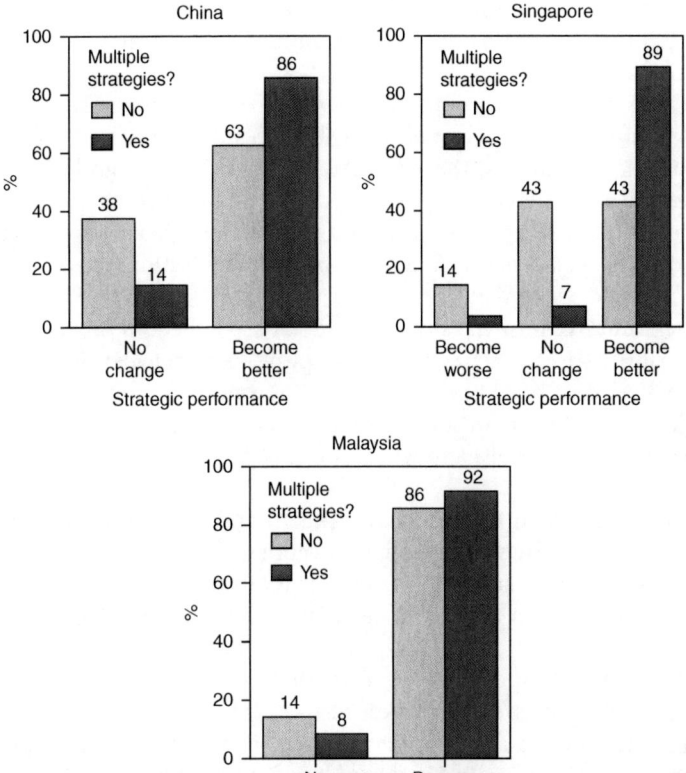

Symmetric measures

Country where based			Value	Approximate significance
China	Nominal by nominal	Phi	0.256	0.168
		Cramer's V	0.256	0.168
	N of valid cases		29	
Singapore	Nominal by nominal	Phi	0.467	0.022
		Cramer's V	0.467	0.022
	N of valid cases		35	
Malaysia	Nominal by nominal	Phi	0.094	0.683
		Cramer's V	0.094	0.683
	N of valid cases		19	

Figure 6.6 Comparisons of strategic performance by strategic groups in selected countries

an organisation uses multiple strategies for the Singapore-based organisations, but this is not true for the China and Malaysia-based organisations.

For the Singapore group, the trend of strategic performance (i.e. the proportion of organisations that became better to the proportion that did not) in the two strategic conditions is significantly different. This significant finding reflects the fact that when multiple business strategies are introduced as a formal strategic process, a big majority (89 per cent) of Singapore-based organisations improved their strategic performance.

When multiple business strategies are not used, the opposite is true (the majority – 57 per cent – had not improved strategic performance while 43 per cent had improved it). This indicates that, in Singapore, it appears that there is an association between the degrees of strategic performance (two levels) and whether or not multiple strategies (two options) were used.

For the China and Malaysia-based organisations, there is no such significant result. In both strategic categories, the majority of firms had improved their strategic performance (86 per cent for yes and 63 per cent for no in China; 92 per cent for yes and 86 per cent for no in Malaysia).

In short, in each individual country, the majority of organisations used multiple strategies. Regardless of strategies pursued, the majority of China and Malaysia-based organisations had a better strategic performance. However, in Singapore, for those employing multiple strategies, the majority had improved their performance, yet when multiple strategies were not used, the majority had not improved their strategic performance at all.

It can be assumed that in Singapore, the type of strategic option provides significant impacts on strategic performance for service organisations: the higher level of strategic performance is associated with multiple strategies and the lower level with non-multiple strategies. The type of strategy, on the other hand, does not influence organisations' strategic performance in China and Malaysia: service organisations can achieve a sound strategic performance no matter what types of strategy are selected.

As too few numbers of cases for some categories (e.g. become worse) were obtained, there is an inevitable limitation on the research results. In this sense, no conclusions can be drawn based upon the observed results. More cases are needed if it is sought to prove the significance of the findings in individual countries.

6.4.2 Cross-national environment–performance relationships

The resulting scatter plot is shown in Figure 6.7. First, there seems to be some general trends in the data such that the higher levels of strategic performance are associated with lower levels of perceived environmental uncertainty. Second, in each individual country, the scatter plot shows that the perception of environmental uncertainty is fairly fragmented. However, the majority of organisations had improved strategic performance (there are very few cases that had strategic performance levels below 3, which means becoming worse). Third, another noticeable trend in these data is that none of the three countries had cases having a strategic performance of becoming much worse (strategic performance below 2).

Amongst the three selected countries, for the China and Malaysia groups, perceived environmental uncertainty was at a relatively low level (mean = 3.6 and 4.0 respectively) yet they achieved a relatively high level of strategic performance (mean = 5.2 and 5.4 respectively).

For the Singapore group, perceived environmental uncertainty was the highest (mean = 4.7) and the Singapore-based organisations had a relatively poorer strategic performance than that of the other two countries' groups (mean = 4.9).

These observed results indicate a pattern that, when making a comparison among individual countries in an East Asia context, higher levels of strategic performance are associated with lower levels of perceived environmental uncertainty.

In order to gain a result from the significance tests, a matrix is displayed giving the Spearman correlation coefficient between the two variables, perceived uncertainty and strategic performance, in each of the three selected countries (Table 6.7).

For China, the marginally significant result indicates that there is an association between perceived uncertainty and strategic performance, $r (29) = -0.323$, $p = 0.087$.

For Singapore, the correlation coefficient between the two variables of uncertainty and performance is $r (35) = -0.375$, $p = 0.027$;

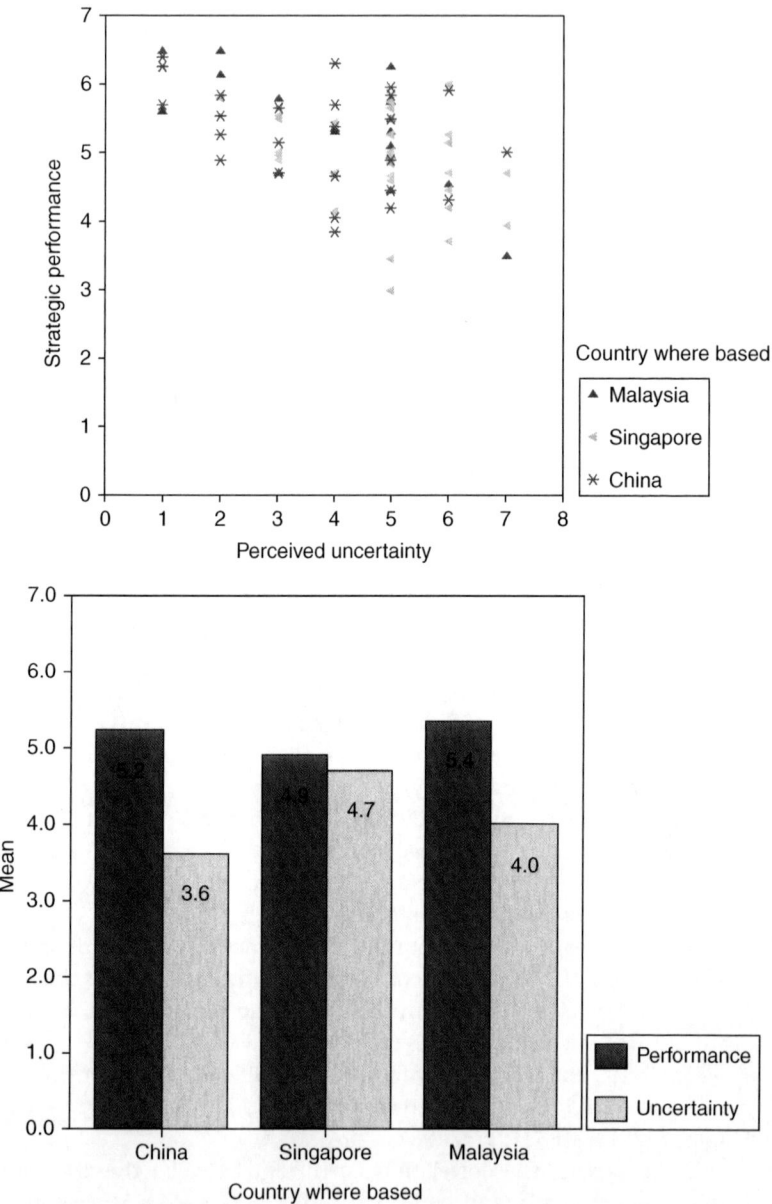

Figure 6.7 Perceived uncertainty–performance relationship

Table 6.7   Perceived uncertainty–performance correlations

Country where based			Strategic performance
China	Correlation coefficient	Perceived	−0.323
	Significance (2-tailed)	Uncertainty	0.087**
	N		29
Singapore	Correlation coefficient	Perceived	−0.375
	Significance (2-tailed)	Uncertainty	0.027*
	N		35
Malaysia	Correlation coefficient	Perceived	−0.629
	Significance (2-tailed)	Uncertainty	0.004*
	N		19

*Correlation is significant at the 0.05 level (2-tailed).
**Correlation is marginally significant at the 0.10 level (2-tailed).

and for Malaysia, the correlation coefficient is r (35) = −0.629, p = 0.004. The significance value for this correlation coefficient is less than 0.05; therefore, in both Malaysia and Singapore, there is a significant relationship between strategic performance and the degree of perceived environmental uncertainty. As all correlations are negative, it can be concluded that as the degree of environmental uncertainty decreases, there is a corresponding improvement in levels for strategic performance.

Furthermore, among the three countries, the association for the China group is similar to the Singapore group but is weaker than the Malaysia group. The Singapore and China groups had a weak correlation coefficient, whereas the Malaysia group showed a strong correlation between strategic performance and environmental uncertainty. This means that, when the degree of perceived environmental uncertainty decreases at the same level, the corresponding improvement of strategic performance is different: the China-based organisations improved in a manner similar to the Singapore organisations; the Malaysia-based organisations show a greater improvement in strategic performance than the China and Singapore organisations.

6.5   Cross-national comparisons: differences

Having compared the differences of correlations in each of the three countries, we now proceed to compare the differences in the levels of perceived environmental uncertainty and strategic performance for organisations employing various strategies.

In order to gauge the situation more precisely, the five categories of business strategies are also merged into two. Balancers and analysers are combined to form a category called 'using multiple strategies'; prospectors, defenders and reactors are combined into the second category of 'not using multiple strategies'.

6.5.1   Difference of uncertainty by business strategies

A box plot graph was displayed (Figure 6.8). When using the median as a measure of central tendency, it is observed that organisations in China using multiple business strategies such as analysers or balancers perceived a lower level of environmental uncertainty (median = 3) than those without multiple business strategies (i.e. prospectors, defenders and reactors, median = 4).

The Malaysia group shows exactly the same trend with respect to the level of environmental uncertainty: the degree of environmental uncertainty perceived by organisations using multiple business strategies is lower (median = 4) than that perceived by organisations not using multiple business strategies (median = 5).

However, the Singapore group shows a different pattern from the China group. Regardless of whether or not multiple business strategies were used, the level of perceived environmental uncertainty is the same (both medians are 5).

When conducting Mann–Whitney tests, none of the differences is significant: for the China group, $p = 0.536$; for the Malaysia group, $p = 0.264$. This means that there is no significant difference of perceived environmental uncertainty between using and not using multiple strategies. In other words, though different strategic options are conducted by organisations, it does not make any difference for the level of perceived uncertainty.

6.5.2   Differences of performance by business strategies

A graph displaying a box plot is presented in Figure 6.9. Using the median as a measure of central tendency, in each of the three

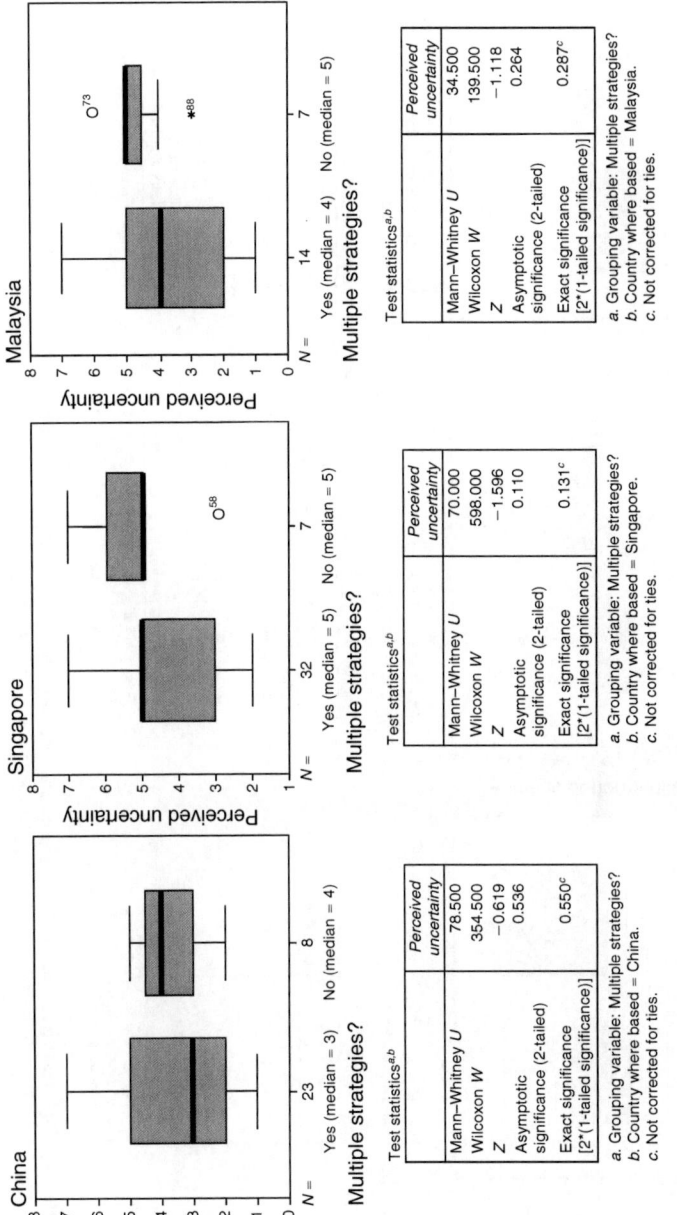

Figure 6.8 Perceived uncertainty–strategy association in China, Singapore and Malaysia

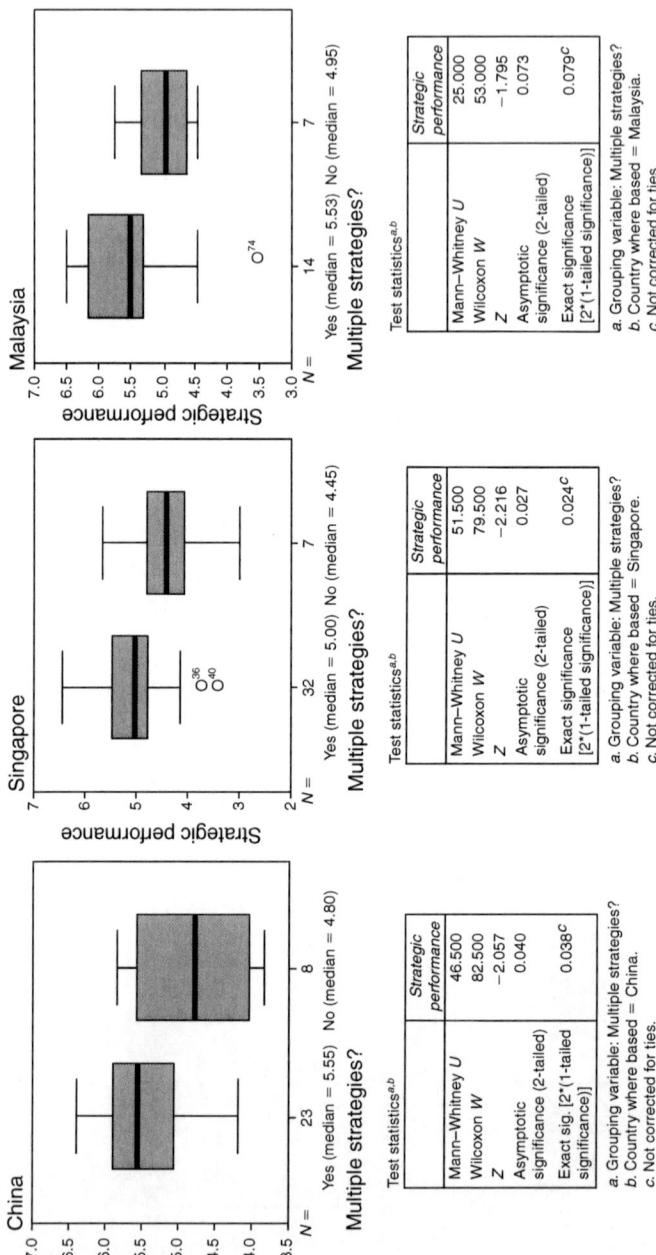

Figure 6.9 Business strategies and associated performance in China, Singapore and Malaysia

countries the common character is that balancers or analysers outperformed organisations with prospector, defender or reactor strategies. In order to examine the significance of these differences, Mann–Whitney tests were carried out.

Obviously, in each of the three countries, the level of strategic performance varied when conducting different strategic business approaches. In China, analysers and balancers significantly outperformed (median = 5.55) organisations with prospector, defender or reactor strategic directions (median = 4.45), $p = 0.04$.

For the Singapore group, organisations using multiple strategies also significantly outperformed (median = 5.00) organisations not using multiple strategies (median = 4.45), $p = 0.027$. In Malaysia, the strategic performance of analysers or balancers was better (median = 5.53) than that of the other three strategic types (median = 4.95), and this difference is marginally significant, $p = 0.073$. These results are consistent with the observation demonstrated above. It can be concluded that appropriate strategic options should essentially be relevant to a better performing business.

6.6 Summary

Regardless of managerial perceptions on the business environment, they may not have a significant impact on the selection of oil service companies' long-term business strategies. Nevertheless, balancers felt relatively more comfortable with the business environment than did prospectors, reactors and analysers. Reactors appeared relatively 'lazy' in monitoring the business environment in which they operated.

Strategic performance is negatively related to perceived environmental uncertainty or hostility: that is, the higher the perceived environmental uncertainty or hostility, the weaker the strategic performance.

Analysers or balancers outperformed defenders and reactors in the East Asian market, suggesting that using either an analyser or balancer strategy can guarantee organisations a better performance than those using a defender or reactor strategy. Having a prospector strategy does not yield an organisation a significantly different strategic performance from analysers or balancers.

Furthermore, a better strategic performance is also associated with differentiation-oriented strategies, while a poorer strategic performance is associated with the competition position of low customer added value premium prices. A differentiation strategy is found to be a more appropriate option than a low-cost strategy for service organisations operating in East Asia. Hybrid organisations may achieve a higher level of strategic performance than low-cost and non-purpose strategic groups.

In each of the three countries, a higher strategic performance is associated with a lower level of perceived environmental uncertainty regardless of the type of strategy; different categories of strategic options do not show differences in the perceptions of perceived environmental uncertainty but do show differences in relation to strategic performance.

In both Malaysia and Singapore, there is a significant relationship between strategic performance and the degree of perceived environmental uncertainty: when the degree of environmental uncertainty decreases, there is a corresponding improvement in levels of strategic performance. When the degree of perceived environmental uncertainty decreases at a same level, the corresponding improvement of strategic performance is different: the China-based organisations improved in a manner similar to the Singapore organisations; the Malaysia-based organisations show a greater improvement in strategic performance than the China and Singapore organisations.

The level of strategic performance is significantly related to whether or not an organisation uses multiple strategies (i.e. balancers or analysers) for the Singapore-based organisations, but this is not true for the China and Malaysia-based organisations. In Singapore, the type of strategic options provides significant impacts on strategic performance for service organisations: the higher level of strategic performance is associated with multiple strategies and the lower level associated with non-multiple strategies. The type of strategy, on the other hand, does not influence organisations' strategic performance in China and Malaysia: service organisations achieved a sound strategic performance no matter what types of strategy were selected. This indicates that the domestic business environment in both countries was at the time in favour of service companies' interests in pursuing prosperous business opportunities.

7
Recent Trends of the Energy Service Industry in East Asia

This chapter discusses the most recent trends of the oil and gas service industry in the light of the findings from the empirical research. It reviews the current dominant business environmental conditions and strategic developments within the oil and gas service sectors. In particular, this chapter highlights the impacts of the global credit crunch on the service industry during the turbulent times of 2008–9. Business strategies deployed by the major domestic service companies are explored to reflect the recent strategic moves in China, Singapore and Malaysia.

7.1 Overview of the business environment in 2002–9

The sustained high oil price environment and strong demand growth over the last few years have resulted in intensification of industry activities (PETRONAS, 2009). PETRONAS, the Malaysian national oil company, finds that, over the period between 2002 and 2008, the oil and gas service industry had benefited from significant increases in the volume of demand and the value of service rates. However, this situation has also stimulated a number of downturn pressures on the service sector.

Overall, within the oil and gas service industry, costs have been driven up even higher than the increase in crude prices. According to PETRONAS (2009), over the period between 2003 and 2008, West Texas Intermediate (WTI) crude prices recorded a cumulative increase of 182 per cent. In comparison, the daily charter rates for drilling rigs increased by almost 300 per cent and the average price of steel increased by 225 per cent per tonne. In general, the increase in costs has resulted in oil and gas companies incurring higher capital expenditures to sustain operations.

In most geographical markets, service companies' profitability was also offset by increasingly higher operating costs. In its 2008 annual report, PETRONAS (2009) suggests that the cost escalation was compounded by the industry's move to harsher operating environments such as deep water, more expensive raw materials such as steel costs, and higher equipment and labour costs.

Market conditions have worsened further due to the lack of engineering and construction capacity. For some service equipment such as floating production units (FPUs), drilling platforms or rigs, and offshore support vessels, it usually takes approximately an average of 12–24 months (i.e. lead time) to complete the construction and installation work. When high oil prices had driven intensified oil and gas industry activities, in some service segments equipment and service supply could not catch up with rapidly growing demand. In addition, there is an acute shortage of experienced and skilled personnel throughout the world petroleum service industry (US Commercial Services, 2009; Singapore EDB, 2009).

The global oil and gas industry environment has become increasingly volatile and uncertain due to the financial and credit crisis commencing in mid-2008. The credit crisis has reduced the availability of liquidity and credit to fund the continuation and expansion of industrial business operations worldwide. The shortage of liquidity and credit coupled with substantial losses in worldwide equity markets during 2008–9 has led to a recession in the United States and Europe, and also resulted in the slowdown of the global economy (PETRONAS, 2009).

Demand for services depends on the levels of oil and gas exploration, development and production activities or the oil companies' expenditure levels. Oil and gas activities and expenditures are directly affected by the trends of oil and natural gas prices. A slowdown in economic activity resulting from the worldwide recession, along with lower prices for oil and gas, would naturally reduce worldwide demand for energy and, in turn, demand for oil and gas services (Pride International, 2009).

Due to the global economic meltdown, it is evident that developments in the external environment can have a significant influence and deep impact on service activities. These external environmental factors include commodity price volatility, foreign exchange movements, the credit crunch, recession, geopolitics and the regulatory environment (Sembawang, 2009; Sembcorp, 2009). Other than affecting service companies' operations, some of the service companies' vendors, contractors, suppliers and customers have also been affected.

To summarise, the impact of the global credit crunch on the oil and gas service industry has appeared in the following forms.

7.1.1 Reduction in demand for services and downturn of service activities

Demand for upstream oil and gas services such as drilling is particularly sensitive to the level of exploration, development and production activity of, and the corresponding capital spending by, oil and natural gas companies. If demand for oil and gas services declines, the service industry could experience a decline in service rates for new contracts and a slowing in the pace of new contract activity (Ensco, 2009; Noble Corporation, 2009). The slowing of upstream activities in turn drives the reduction of the service pace for the oil and gas industry.

Crude oil prices declined significantly in the second half of 2008 and the forecast crude oil prices for the remainder of 2009 are not expected to return to prior levels (PETRONAS, 2009; CNOOC, 2009). Regional oil majors such as CNOOC and PETRONAS believe that there is no assurance that the financial crisis will improve. As a result, the impact of the current global economic recession on service companies' future liquidity and financial conditions cannot be predicted.

Consequently, after the sharp correction in oil prices and the equities slide, the year 2009 has been commonly viewed as a difficult time for the world petroleum industry. Many service segments have been badly hit by the volatile petroleum economy as there has been a reduction in many international oil companies' E&P activities, capex spending, as well as drilling utilisation and rates. A general sign shows that, compounded by job cuts, some market segments have become tighter, at least for the short term, for goods, equipment and services. According to the China Ship Industry Association (2009), service suppliers such as CSIC and CSIC saw a significant decrease in new orders for 2009.

7.1.2   Tightening credit market conditions' adverse impact on all oil and gas participants

The credit crisis could affect service companies' customers, causing them to fail to meet their obligations to the service contractors (Ensco, 2009). Any prolonged reduction in oil and natural gas prices or material contraction of the petroleum customers' cash flow or liquidity, including their access to capital, could result in lower levels of exploration, development and production activity (PETRONAS, 2009).

On the back of the global financial crisis, some oil companies have not been able to sustain their activities due to cost escalation and difficulties in financing, and many projects have been deferred during the 2008–9 period (PETRONAS, 2009).

Some contractors (COSL, 2009; CSSC, 2009) find that, in order to minimise their own risks, oil companies may:

• cancel their contracts due to a reduction in their exploration, development or production programmes;
• or try to avoid or reduce their obligations to service companies under previous agreements or contract conditions;
• or seek to obtain a comparable supplier at a lower price.

Drilling contractors such as Noble Corporation (2009) contend that lower levels of petroleum industry activity could result in a corresponding decline in the demand for upstream-related services. The decline in upstream service demand could in turn have a material adverse effect on oil service companies' overall performance.

The financial crisis also hit the ability of subcontractors to meet scheduled deliveries of their service goods for major service companies. This could, in turn, damage the reputation of upstream service contractors as they may fail to fulfil their contract obligations to oil clients.

7.1.3   Smaller-sized service contractors' struggle to maintain activities

The credit crunch and related instability in the global financial system may have impacts on service companies' liquidity and financial condition. Since banks and other lenders have suffered significant losses in the equity market, standards for lending are becoming stricter. There is a general restriction on the availability of credit. It may be difficult or more expensive for most service companies to access the capital markets or borrow money whenever they would like, or need, to access capital. This situation has an adverse impact on their ability to react to changing economic and business conditions (Ensco, 2009; Pride International, 2009; Noble Corporation, 2009). Service companies have had to postpone some of their desired operations, capital expenditure, and proposed acquisitions. As the financial market tightens, some service companies such as FPU builders or drilling contractors are finding it difficult to renew their fleet by new vessel construction and conversion projects (Sembcorp, 2009).

7.2 Recent trends of business conditions in East Asia

The overall growth of the service market throughout East Asia has remained strong despite the 2008–9 global financial problems compounded by the sharp downturn in oil prices. The growth has been partially sustained by growing exploration, development and production activities as well as an increasing demand for energy in the region.

7.2.1 Economic impact

China's rapid economic development created a strong demand for energy with its compound annual growth rate (CAGR) of 6.5 per cent between 2001 and 2007 (National Bureau of Statistics of China, 2007), while the country meets 50 per cent of its crude oil demand with imports. With demand being maintained at high levels, investment in oil and gas exploration will be sustained (COSL, 2009).

As the Chinese petroleum industry is largely dependent on domestic demand and most related products/services have been provided for the Chinese market, the global financial crunch since mid-2008 has had limited impact on the Chinese petroleum and energy service industries. Despite new orders from abroad dropping significantly, the Chinese service sector is still continuing normal progress in expanding offshore engineering capacity in Qindao and Zhuhai (CNPC, 2009; CNOOC, 2009).

According to COSL (2009), exploration in Southeast Asian and PRC waters will remain active; major oil clients operating in the region are keeping their capex for 2009 intact or slightly higher than previous levels. New discoveries in Bohai Bay will also take drilling and other activities offshore China to new heights in the years to come. Service companies in China stand to benefit from a growing demand for oil-field services.

CNOOC intends to maintain investment in its major businesses over the period between 2009 and 2012. It will also accelerate its pace of investing in deepwater exploration and increasing workload in this area (COSL, 2009). Following its announcement of production start-up for two new oil projects, Bozhong 28-2S and Panyu 30-1, CNOOC plans to bring six other offshore oilfields on stream in 2009. It has conducted conceptual studies on at least two new projects in Bohai Bay, Jinzhou 25-1 and Kenli 3-2 (CNOOC, 2009).

According to Baker Hughes (2009), demand for offshore drilling capacity remains stable in China and Malaysia. The number of drilling rigs in use in China and Malaysia has maintained a growth trend from 2002 to 2009. Average rig count for 2009 is still relatively high for China and Malaysia, up from around 21 and 13 units chartered in 2008 to 23 and 14 units respectively. Day rates in these regional markets remain stable due to strong demand and geographical reasons (COSL, 2009).

High oil and gas prices and an ageing offshore fleet have also buoyed the marine and offshore industry in Singapore (Singapore EDB, 2009). The business environment remains favourable for the service sector in this country (US Commercial Services, 2009).

For some export-dependent service segments, the regional near-term market is tightening as there have been contract cancellations, postponed projects, delays of delivery of service equipment, and changes in contract terms and conditions (CSSC, 2009; CSIC, 2009). In some market segments, companies could see a reduction in demand for their services (COSL, 2009). Some petroleum clients are waiting for further funding injection, or otherwise are having to put back their proposed development activities (MSIC, 2009).

7.2.2   Political/regulatory impact

Unlike anywhere else in the world, the business environment for the service sector is relatively stable due to government support. In China, the negative impact of the global credit crisis has been counterbalanced by government incentive policies.

Chinese policymakers have set up short- and medium-term policy objectives to help indigenous entities overcome the challenge of the global financial crisis. The aim is to help state-owned service firms to carry on smoothly with the construction, conversion or upgrades of equipment and facilities. Funding from the government is available to help firms finance their projected investment programmes including new build construction, upgrades, refurbishment and repair. Government funds also ensure normal business activities (China Ship Industry Association, 2009).

Beijing decided to pump CNY 4 trillion into the economy in the form of an economic stimulus package including large investment in fixed infrastructure, some of which has been used for the construction of offshore engineering and manufacturing bases in Qindao and Zhuhai (CNPC, 2009; CNOOC, 2009).

Singapore has retained its strength in oil and gas service businesses as its offshore industry has benefited from government support. As the global market scales up, the Singaporean government plans to enhance its competitive advantages in maintaining a world-class capacity of storage infrastructure and the presence of first-rate financial institutions. In order to reinforce its leading position as an international hub for the petroleum refinery, trading and logistics businesses, the government is giving strong support to investment and resources for its service industry. Its key domestic service sectors comprise shipyards and entities involved in offshore or marine engineering-related business (Singapore EDB, 2009). In addition, the Malaysian government has provided support to help local enterprises overcome financial difficulties and maintain business as normal.

7.2.3 Impact of technology

Today, technology is pushing out the smaller operators as it becomes more and more expensive to compete with the 'big brothers'. Technology in the oilfield usually comes with fairly hefty price tag (Croy, 2000). Both national and international contractors provide services to oil clients in offshore China and Malaysia. Local contractors normally offer high-labour low-technology services such as small-scale fabrication, including offshore structures. Foreign contractors offer low-labour high-technology services to support operations, subsea inspection, engineering machinery overhaul, etc. In this regard, Chinese and Malaysian contractors are less competitive compared with international organisations due to their relatively less advanced knowledge of offshore technologies.

E&P companies have moved towards the harder-to-find new hydrocarbon reserves, which are located mainly in deeper waters, harsher climatic conditions and environmentally sensitive regions. Gaining access to such reserves is more difficult, riskier and technologically more challenging (PETRONAS, 2009). Service companies with expertise in such fields will find themselves well positioned to find prosperous business opportunities.

7.2.4   Impact of customers

Like anywhere else, due to the downturn of the global petroleum industry, some international oil companies have reduced their E&P activities and put pressure on the supply and service sector. As a result, the supply and service industry in East Asia is affected by a number of downside factors from oil clients.

Oil companies may have asked to cancel their contracts or letter of intent or agreements; or they might request to change contract or agreement details such as resetting delivery schedule or product type; or have requirements relating to cutting service prices and payment delays.

The Chinese shipbuilding industry was hit badly due to worsening demand conditions. According to the China Ship Industry Association (2009), by the end of 2008 there were very limited new orders for the major Chinese shipbuilding and subsea equipment service companies such as Shanghai Waigaoqiao Shipbuilding Company (SWS) and Dalian Shipbuilding Industry Corporation (DSIC). Some withdrew contracts, of which most are long-term delivery orders. In 2009, the priority for the offshore shipbuilding/equipment industry in China has shifted from taking new orders to securing the smooth execution of existing ones.

Despite the adverse impact on short-term exploration and development activities, China's long-term energy development strategy remains the same. Based on its energy demand, China is accelerating offshore exploration and development programmes (COHC, 2009).

CNOOC (2009) announced a total of approximately USD 29 billion investment in deepwater oil and gas exploration programmes in the South China Sea. As the biggest development plan for offshore China, this investment will continue for more than ten years. CNOOC also has 13 other investment plans for the oil and gas development programmes which are expected to commence by 2010 offshore Bohai Bay.

Apart from CNOOC, CNPC and Sinopec have also accelerated their deepwater offshore exploration and development activities. The two companies have also been granted permission to explore blocks in the South China Sea by the Chinese government. Deepwater development is now listed as an important project by these two Chinese oil giants. As a result, some service segments such as drilling and helicopter services will benefit from oil companies' deepwater expansion in China and Malaysia. Shallow-water services will also see an increase in the Chinese offshore service industry as there will be an increasing demand from the three big national oil operators (COHC, 2009).

7.2.5 Emerging opportunities and challenges in the service sector

A combination of various factors has made the business environment for service companies more uncertain. Faced with a sharp drop in oil prices, volatile currency exchange rates and a short-term declining market demand in some segments (COHC, 2009), there are both emerging opportunities and challenges in the service sector.

Opportunities for service companies in China include deepwater developments and other onshore and near shore developments (COHC, 2009). PETRONAS's new developments in deepwater E&P in Malaysia have also opened up new opportunities for growth.

The US Commercial Services (2009) report that, in Malaysia and Singapore, there are opportunities for international companies specialising in:

- offshore engineering,
- construction and contracting services or project consultancies,
- drilling,
- offshore platform and rig fabrications and operations,
- shipbuilding and marine engineering,
- transportation equipment.

Given the increased emphasis on gas in China and Southeast Asia, pipeline technology has also been identified as another growth area for service companies. There are also growing demands from some niche market segments in the region, including geographical surveys, navigation and positioning, hydrographic surveys and underwater inspection services (US Commercial Services, 2009).

In addition to the above, opportunities are available for service companies to improve their competitive advantages such as expanding service or equipment capacity. Given the global economic downturn, it is possible that some of their peer competitors' assets such as rigs being built speculatively could become available for purchase (COSL, 2009).

In contrast with these opportunities, some organisations (COHC, 2009; PETROBRAS, 2009; Singapore EBD, 2009) also highlight major challenges confronting the service sector in East Asia:

- There are increasing raw material costs and labour costs.
- Fluctuating exchange and interest rates have an adverse impact on the operating performance of service companies.
- Local SMEs need to continuously develop new and innovative solutions to meet the growing demand for deeper water technology and high-tech equipment.
- Domestic SMEs are not exposed to opportunities for collaboration, where they could combine strengths and resources to secure larger projects. Most of the SMEs, being subcontractors, provide services to only one or a few customers.
- In some market segments, there is a shortage of skilled manpower and the oil and gas industry has to rely on labour from elsewhere.

7.3   Current strategic orientations of service companies

Regardless of business types or associated locations, oil and gas service businesses in East Asia fall into four major streams depending on the strategy applied: defender, prospector, analyser and balancer. Although these forms of strategy were introduced in Chapter 3, the following will summarise again the major characteristics of these strategies.

As a defender in East Asia, a business may pursue concentrated growth across the entire oil and gas industry, with services to companies from exploration and appraisal, to development and production, and to decommissioning; and they may operate across different regional markets, home and abroad.

Prospectors pursue market development by finding new markets for their existing products and services. They are also active in product/service development through a focus on differentiation and therefore new products/services can be available for existing clients. An analyser may have concentric or conglomerate business units. In addition to a mixed feature of combining being a defender and prospector, it also benchmarks changes in identified competitors' businesses and seeks to obtain potential profitable or new technologies from oil companies, technological institutions or government parties.

A balancer is a combination of the above three types of strategies. It may contain businesses operating in a non-oil and gas sector.

The remainder of this section will use examples drawn from the leading domestic or regional players to depict existing strategic trends.

7.3.1 Defender strategy

Adopting a defender strategy has become the backbone of organisational behaviour for service businesses operating in East Asia. Very commonly, a defender is recognisable by its capacity expansion, improvement of managerial capabilities and operational efficiency, as well as market development. A defensive approach could at least provide a defender with the stability of maximising business revenue.

In recent years, actions have been taken by service companies in the improvement of production capabilities. Companies such as COOEC, CPOE, CSSC and CSIC seek to expand their production capacity through new constructions of their manufacturing bases.

A defender seeks to strengthen its competitive position in existing markets. This may be achieved by improving its focus on customer needs and reducing costs while maintaining the required standards for safety and reliability. COHC, a major Chinese offshore helicopter service supplier, has adopted such a strategy.

In China, major state-owned service companies continuously improve and streamline their work processes to maximum operational and cost efficiency and the utilisation of resources. Sembawang Shipyard, a Singaporean ship repair and conversion company, also carried out activities to improve operational efficiency through production process improvement.

Additionally, some domestic service companies try to achieve capacity expansion and operational efficiency through taking advantage and learning the experience of their international joint venture partners. For example, COOEC and SeaMetric, a Norwegian offshore heavy lift service contractor, proposed setting up a joint company in which SeaMetric will provide financial resources, operational management and specialist engineering expertise while COOEC will provide crew and engineering support.

For larger service organisations, production and managerial efficiency improvement is also seen to be associated with the trends of industry consolidation and restructuring. In the Chinese shipbuilding industry, a number of subsidiaries of CSSC or CSIC have merged so that they can achieve their desired economies of scale. Today's CSSC is an outcome of a series of industrial restructuring and organisational rebranding processes.

Elsewhere in Malaysia, following the incorporation process of MISC and MSE, and PETRONAS's strategic taking over of MISC, with a much more extended capacity, Malaysia Marine and Heavy Engineering Sdn Bhd (MMHS) is also a rebranded name of MSE and a wholly owned subsidiary of MISC (MISC, 2009).

With regard to market development, a defender focuses on entering into new long-term contracts for its existing products/services and obtaining renewal of existing contracts.

One popular approach is that a defender constantly seeks to enhance its existing customer relationships so that use of its existing products or services can be increased. COSL has established long-term strategic relationships with CNOOC and has signed or renewed contracts for the next two years with other international oil clients. In order to expand its business and regional presence, KBR, an international engineering and construction service company, leverages its partnerships and alliances with local offshore fabrication yards based in Singapore.

Given these interdependent relationships, service firms can take steps to minimise the adverse impacts of the global financial crisis while maintaining their own interests. This is evidenced by Sembawang's mutually agreeable payment deferment arrangements with customers that are facing financial difficulties. Meanwhile, the company also monitors its exposure to its customers, vendors and other counterparties such as banks, and engages with them on a regular basis.

7.3.2 Prospector strategy

Some service companies become key players in the industry by means of a prospector strategy. Being a prospector, firms are always associated with in-house activities in innovation and proprietary designs. In order to maintain its competitive edge and relevance in the fast-changing offshore and marine market, Sembawang, for instance, pursues a strategy of continuous investment in research and development to enhance its proprietary engineering and design. A prospective strategy helps a company to improve its business performance. As a result of being innovative, Sembawang saw operating margins increase from 5.9 per cent in 2005 to 9.9 per cent in 2008 (Sembawang, 2009).

A prospector is also characterised by its intention to capitalise on its broad geographic coverage, long-term customer relationships and its asset capabilities in pursuing new opportunities in new and/or fast-growing markets. Following in the footsteps of globalisation of CNOOC, their parent company, COOEC and COSL are seeking opportunities to expand overseas. More details on these two companies' prospective characteristics will be given in the next two sections.

7.3.3 Analyser strategy

An analyser stands somewhere between a defender and prospector by monitoring and following changes initiated by others. It combines characteristics of both defender and prospector. Quite often, it is in partnership with various technology partners from the government, educational and private sectors for research and development projects. By doing so, an analyser can monitor and follow closely any changes in the industry.

A representative analyser example shown below is the strategy adopted by COOEC. COOEC's strategic aim is to enhance its business presence in the domestic and overseas market, and become a leader in the offshore service industry in China and Southeast Asia. It intends to position itself as a pioneer subsea engineering service provider through enhanced core competence, the development of technological advances, and the improvement of managerial and operational efficiency.

Coupled with its core business in offshore oil and gas field construction and maintenance, COOEC is gradually deploying a so-called 'two wings' system to generate a wide range of business units. The company has a combination of various business units including design, construction, installation, IMR (inspection, maintenance and repair) and scientific research. According to COOEC, its 'left wing' business means serving the downstream market and the 'right wing' means the services provided for the deepwater market. Throughout its major business domain, COOEC monitors international standards so as to facilitate mutual infiltration of technology and management in the worldwide offshore service market.

7.3.4   Balancer strategy

A balancer is usually seen to be well positioned in its marketplace and appears to be a market leader in the industry. In addition to the characteristics of operational efficiency, capability of initiating industrial changes and benchmarking successful industry practices, a balancer also pursues profitable new businesses beyond the oil and gas sector.

A typical feature marking out a balancer is its business units with a wide range of products, equipment and services. COHC, for instance, serves the offshore industry and also competes in the emerging medical service (EMS) and search and rescue (SAR) sectors. Its competitive advantage stemmed from its available capacity with a good range in its helicopter fleet. Its key strength is attributed to its ability to operate twin-engine medium and heavy helicopters with highly trained pilots in complex situations. Typically, customers in its target market require the service provider to meet stringent quality standards on a long-term basis.

COSL is situated as the number one oilfield service provider offshore China. It owns and operates the largest and most diverse fleet of rigs and support vessels offshore China. Its supply capability has provided the company with economies of scale and the capacity to service the entire offshore China market.

For COSL's well services and marine support and transport services, the business unit has a capacity of integrated products with an enhanced logging imaging system (ELIS) and formation characteristics tools (FCT). As of 2009, its capacity has been expanded with the addition of two new utility platforms and the introduction of new equipment; 21 newly delivered working vessels, and 5 under lease.

Credited to its comprehensive capacity availability, new orders from this business segment have continued to increase at stable prices and revenues generated from the segment's overseas market also continue to grow.

COSL has integrated offshore oilfield services across exploration, development and production activities. The integrated service lines allow the company to provide strategic and comprehensive coverage of its customers' offshore service needs. Most importantly, it allows the company to offer customers reservoir expertise and exploit synergies across their service lines (COSL, 2009).

As a strategic balancer, COSL has a combination of the following characteristics: cost leadership, integrated services, international penetration, and technological innovation. Because of its geographic concentration, lower labour costs and scale of operations, COSL is able to offer its services under a more competitive cost structure than many international companies offshore China. Its competitive cost structure provides the company with a strong platform for expansion into international markets.

In addition to its dominant position in the domestic market, COSL has implemented a 'Go Out' policy to expand constantly in the international market. The company has also made a technology breakthrough that helped to gain its first equipment construction contract for a 5000 tonne large-scale crane from a foreign client. It also succeeded in undertaking the main body construction of a large-scale nickel smelter which is the first integrated and modularisation type in the world.

In recent years, a strategic trend has emerged that service companies seek to enlarge the range of their businesses through acquisitions (COSL, 2009). An individual transaction for specific mobile offshore drilling units or a transaction for an entire company has been seen as a common practice in the oil and gas industry.

Larger service companies look for additional acquisition opportunities to further strengthen their position in existing markets and expand into new markets. It is evident that some Chinese service organisations are becoming acquisitive and have carried out acquisitions abroad. A representative example is COSL's acquisition of a Norwegian company, Awilco Offshore ASA, in September 2008. The successful completion of this acquisition marked a crucial milestone in COSL's business expansion into its international markets.

A further innovative strategic approach to becoming a well-established balancer has emerged in the petroleum industry in Malaysia. In January 2008, the offshore logistics arm of PETRONAS, MISC, announced a reverse takeover of Ramunia Holdings Berhad (Ramunia) by disposal of its entire equity interest in MMHE to Ramunia. The purpose is to benefit from expanded fabrication yard capacity and shared expertise and resources.

7.4 Summary

It has been challenging for many oil and gas service companies throughout the world to sustain their position over the turbulent times of 2008–9. This is set to amplify the level of uncertainty in the already complex and volatile oil and gas service industry. It also further intensifies competitive activities in an environment in which it is increasingly difficult to gain new contract awards and there is limited access to financial resources.

In addition to an acute shortage of experienced personnel and equipment, oil and gas service companies will continue to operate in a highly challenging economic environment where escalating costs have eclipsed gains from high rates. Service companies will face significant challenges if conditions in the financial markets do not improve. Smaller-sized service companies may find it harder to survive during such a volatile period.

However, given the new discoveries offshore China and deepwater development offshore Malaysia, the service sector is expected to grow in the coming few years.

Thanks to government support and steady demand, the business environment in China, Singapore and Malaysia remains benign for the energy service industry. Especially, the enduring reform in the Chinese financial sector will ensure state-owned energy service companies have sufficient financial resources to fund their capex spending and business activities.

E&P companies have moved towards new harder-to-find hydrocarbon reserves, which are located mainly in deeper waters, harsher climatic conditions and environmentally sensitive regions in China and Malaysia. Gaining access to such reserves is more difficult, riskier and technologically more challenging. Service companies with expertise in such fields will find themselves well positioned to find prosperous business opportunities throughout the region.

Due to the downturn in the global petroleum industry, some international oil companies have reduced their E&P activities and put pressure on the supply and service sector. As a result, the supply and service industry in East Asia has been affected by a number of downside factors including: contract cancellations, requests for altering contracts or agreements, cutting service rates, and payment delays.

Regardless of the world economic downturn and uncertain market conditions, service businesses in East Asia can be clustered into four major strategic streams: defensive, prospective, analytical and balanced.

Adopting a defender strategy has become the backbone of organisational behaviour in the service sector. A defender is characterised mainly by capacity expansion, operational efficiency improvement and market development. Some service companies become key players in the industry through a prospector strategy. Being a prospector, firms focus on improving business competency through innovation and proprietary designs; meanwhile, they also provide a broad range of products and services.

An analyser's characteristics are marked by organic growth, technology-driven activity and overseas expansion. In addition to its operational efficiency, a balancer has the capability of initiating industrial changes and closely following good industry practices. It also pursues profitable new businesses beyond the oil and gas sector.

8
Conclusion

The concluding chapter summarises the results of the empirical research and discussions on the recent trends of the oil and gas service industry. In particular, it provides the impact of the global economic meltdown on the service sector, and compares empirical research results and major trends generated in recent years. Verifications of the research propositions are illustrated. Implications of the research findings for organisations operating in East Asia are also explored.

8.1 The objective business environment

8.1.1 Generic trends of the service sector in East Asia

The energy service sector is characterised by a large number of companies, which are widely diverse in terms of size, product or service provided and niche activities. Oil and gas service companies have developed very quickly and formed the major part of the supply chain within the petroleum industry.

Service companies range from small to very large indigenous and international firms, including technology- and labour-intensive enterprises. Some firms only focus on certain areas while others have vertically integrated businesses. Many service companies (although highly specialised) provide services not only to the oil and gas industry, but also to other energy industries like electricity and nuclear, or even provide services outside the energy industry.

In spite of the 2008–9 volatility of the global petroleum industry and the 1997–2000 economic uncertainty throughout the Asia-Pacific region, the business environment in China has been benign for oil and gas service activities. Some sectors such as the drilling industry in China are examples of service companies operating in a favourable industrial environment.

Reform of the Chinese petroleum industry has created more room for companies to have greater control of resources and more freedom to manage their own affairs. Consequently, development of the service sector has become more market-oriented. This also represents a considerable opportunity for both domestic and foreign service companies to serve the Chinese market or oil clients.

As compared to China, Singapore and Malaysia appeared more vulnerable when confronting the regional financial crisis and global economic meltdown.

The maturity of the domestic resource base in East Asia made overseas investment attractive. Although the oil service firms heavily depend on oil companies, they do not necessarily appear to be in the same place as their clients. The trend of such a development is to provide services globally. Many multinational firms like Halliburton, Schlumberger or Baker Hughes have set up subsidiaries or joint ventures or offices in the Asian markets like China, Singapore and Malaysia.

The service industry is seen as dynamic. Consolidation among international suppliers has changed the number of service firms in the region of East Asia. The service firms in China used to be owned by the state companies. Now the ownership structure is changing through joint ventures (JVs), joint development, wholly foreign-owned and domestic ownership shifts.

The service market is also changing as it has been consistently restructured. The oil and gas service market is quite different from oil and gas markets. Oil and gas companies focus on production and integrate into midstream and downstream activities, while service companies produce products or technologies, or provide relevant support.

The oil and gas-related political/regulatory factors or localisation policies may be hostile towards service activities and therefore may have an adverse impact on the business opportunities of service companies, especially internationals.

Service activities rely largely on oil and gas clients for their survival. In this buyer–seller relationship, oil companies usually have the dominant power. Historically, state-owned companies in China (e.g. CNPC, Sinopec and CNOOC) and Malaysia (e.g. PETRONAS Carigali) have been the key buyers from the service industry there. In recent years, with more and more international oil firms stepping into E&P activities in the East Asian domestic markets, the buyer structure has been widened. Buyers from the wider energy or marine industries are becoming one of the market segments on the service companies' customer profiles.

Companies in the oil and gas service industry create and maintain competitive advantages with four predominant approaches:

- attaining a competitive price;
- establishing high added value for clients by developing expertise in a particular area;
- applying advanced technologies;
- and developing new technologies through research and development (R&D).

For firms operating in China, the competitive advantages may refer to the advantageous geographical conditions, powerful shareholder background, and first-class processing equipment and technical talents.

In some of the service segments, competition is largely based on pricing. Competitors always imitate what industry leaders have done. However, if anyone can do what leaders are capable of, but cheaper, they would have to face a great financial risk because escalating costs have eclipsed gains from high rates. In China, oil operators take both price and previous quality performance into consideration when they evaluate bidding proposals. Such buyer power has forced significant adjustments in the way that the Chinese organisations operate.

Technology is pushing out the smaller service players as it becomes more and more expensive to compete with the 'big brothers'. Technology in deep water usually comes with fairly hefty price tag. Both national and international contractors provide services to oil clients in offshore China. Local contractors normally offer high-labour low-technology services such as small-scale fabrication, including offshore structures. Foreign contractors offer low-labour high-technology services to support operations, subsea inspection, engineering machinery overhaul, etc. In this regard, domestic contractors are less competitive compared with international organisations due to their relatively less advanced knowledge of offshore technologies.

Today, more and more E&P activities are being carried out in the harder-to-find areas for hydrocarbon reserves, which are located mainly in deeper waters, harsher climatic conditions and environmentally sensitive regions. It has been increasingly more difficult and riskier to gain access to such reserves. Technology developments have also become more challenging. Service companies with expertise in such fields will find themselves well positioned to find prosperous business opportunities in China, Singapore and Malaysia.

8.1.2   The impact of the global credit crunch

Over the period in the second half of 2008, a sharp correction in oil prices and the credit crunch hit the global economy, resulting in a worldwide economic recession and in turn a general slowdown in oil and gas E&P activities. The already sophisticated oil and gas service industry became more uncertain during 2008–9. Many oil and gas service companies found it difficult to sustain their position in such a volatile environment. In addition to existing industry rivalries, and an acute shortage of skilled personnel and equipment supporting offshore activities, it is increasingly difficult for service suppliers to gain new contracts and/or get easy access to financial resources.

Service companies worldwide will face significant challenges if conditions in the financial markets do not improve. Smaller-sized service companies may find it harder to survive during such a volatile period. However, offshore China and offshore Malaysia have retained their activity levels thanks to CNOOC's new discoveries announced in 2009 and the deepwater programme in Malaysia. The service sector in both countries is expected to continue its growth in the foreseeable future.

In addition to a steady demand fuelled by the growing regional oil economy, government support in China, Singapore and Malaysia has made the business environment benign for the energy service companies operating in these three countries. The Chinese financial sector provides special incentive programmes that will ensure state-owned energy service companies have sufficient financial resources to fund their capex spending and business activities. Governments in Malaysia and Singapore have also provided support to help local enterprises deal with financial difficulties and maintain business as normal.

The business environment in China, Malaysia and Singapore has so far been affected by a cluster of combined factors, including a sharp drop in oil prices, volatile currency exchange rates, and a short-term downturn in demand in some market segments. Service suppliers may see opportunities and challenges simultaneously.

A short-term adverse impact has come mainly from petroleum clients. As a result of the 2008–9 global economic recession, some international oil companies have reduced their E&P activities and put pressure on the supply and service sector. The supply and service industry in East Asia has been affected by a number of downside factors: contract cancellations, requests for altering contracts or agreements, cutting of service rates, and payment delays. Nevertheless, from a long-term aspect, the business environment remains favourable for the service sector in China, Malaysia and Singapore.

Service businesses in East Asia can be clustered into four major strategic streams: defensive, prospective, analytical and balanced. As a strategic backbone for a service company operating in East Asia, a defender concentrates mainly on capacity expansion, operational efficiency improvement and market development. A prospector provides a wide range of products and services and is keen on improving business competency through innovation and proprietary design. An analyser combines strengths of both a defender and a prospector. Very commonly, it monitors and imitates changes initiated by its peer competitors.

A balancer appears to have the best business success and is usually seen as a market leader in the service industry. Such service companies have a defender's strengths such as operational efficiency, and have the capability of initiating industrial changes and following good industry practices closely. A balancer is also an active player beyond the oil and gas sector.

8.2 General findings of the empirical research

According to the primary research results, the participating service organisations were divided into independent entities, divisional or subsidiary companies, and operating or business units.

A notable finding shows that most of divisional or subsidiary organisations and most of operating or business units were from Singapore, whereas most of independent organisations were from China. As many of the divisional companies and subsidiaries in Singapore were foreign owned, it proves Singapore is a favourable place and/or a preferred hub for international service organisations to establish their regional headquarters.

The results also show that offshore service organisations were older in Singapore than in China, indicating a longer industrial development of this small and prosperous country. Joint ventures were more popular in China's oil and gas service sector than in both Malaysia and Singapore.

Offshore oil and gas service organisations served not only the oil and gas industry, but also other energy industries like hydropower, electricity and nuclear. In this sense, such enterprises are also regarded as energy service organisations.

The upstream services comprised support activities throughout the four industrial sectors of exploration, appraisal, development and production; some even involved serving other oil and gas industrial sectors away from their upstream activities.

8.3   The perceived business environment

8.3.1   Managerial perceptions on environmental factors

In this study, the six task environmental sectors, namely, economics, technology, regulatory, customers, competitors and suppliers, were considered by executives as important for the growth of their businesses in the region. Among these sectors, economic, technological, customers' and competitors' factors had a relatively strong impact, whereas the regulatory factors were likely to have a moderate impact on the business growth in the region. The suppliers' impact was more likely to be weak. In particular, buyers' power was perceived to be very strong for any service firm. Buyers are dominant in the industry's buyer–seller relationship mainly because of the large number of suppliers throughout the entire oil and gas industry.

The above findings hint that executives within the oil and gas service sector in East Asia should pay close attention to four task environmental sectors: customers, economic, technology and competitors. These key indicators should be assessed when formulating organisational strategies.

8.3.2 Perceived environmental uncertainty

Overall, perceived environmental uncertainty in China was low and in Malaysia, it tended to be at a neutral point. Compared with China, the level of perceived business environmental uncertainty in Singapore tends to be higher. This was related to the following:

- As the number of oil and gas clients in China was larger than in Singapore, the competition in Singapore could be stronger than that in China.
- As there were no oil and gas production operations in Singapore, the market demand for oil and gas services was considered limited as compared to China.
- Competitive actions adopted by firms within the service sector were more aggressive, in terms of price war and imitating competitors, in Singapore than in China.
- Customer demand for new products or services, changes in oil and gas exploration and production levels, changes in well counts and in rig counts were more unpredictable in Singapore than in China. It is because Singapore-based service companies serve mostly non-domestic oil clients worldwide.

Despite the fact that the business environment in Singapore and Malaysia was relatively abrasive, access to available technologies was easier in these two countries.

The offshore oil and gas industry business environment in the South China Sea in 2000 teetered between stable and unstable due to the small market, which, at the time, consisted of five FPSOs, eight platforms and four subsea wells. There were only five oil companies and the scale of service business to the oil industry was not large.

Stable means that each oilfield operator adhered to an annual expenditure plan. For example, flight services for crew change, supply boats for weekly cargo transportation, diesel for fuel, catering for offshore personnel. This makes it possible to forecast the market. Each service firm could discover an oil company's next annual plan through a *guanxi* (personal relationship) and accordingly, make its plans to gain business.

Unstable refers to the need to spend extra on emergency repairs which occurred every year. This kind of expenditure is much higher than spending that has been budgeted. For instance, although in 2000 CACT (CNOOC, Agip, Chevron and Texaco Operators Group) budgeted USD 1.2 million for maintenance, the extra expenditure for that year was over USD 6.0 million for the repair of a 12-inch subsea pipeline damaged by fishing nets (Tan, 2000).

8.3.3 Perceived complexity, hostility and unpredictability

Managerial perceptions on their operating business environment tend to be benign. However, the business environment for the oil and gas service sector in East Asia was perceived to be complex and dynamic.

The service sector was perceived as complex in China, Singapore and Malaysia. The service sector used to be simple but is becoming more sophisticated as it has taken on more and more responsibilities from the oil and gas companies. Many service companies have been set up, of which a large number are very specialised. The increasing number of suppliers makes the service sector more multifaceted. In addition, the causes and effects of the industry are complex.

Industry unpredictability was mainly because of oil price instability. Historically, the oil and gas industry has survived the ups and down of the global economy, stock market crashes, fuel crises, etc., but it always comes back on top. In addition to oil prices, there are so many influences in the industry that it is always in a state of flux, but it always seems to get back on an even keel (Croy, 2000).

With regard to the perceived hostility, barriers to entry into the service industry are high because it requires special knowledge, relevant experience and up-to-date information. Many service firms are obviously always on the lookout for rivals, but existing suppliers are normally so entrenched in their area that any competitor would find it difficult to enter the arena and challenge existing ones. The main barrier would be money! High obstacles for new entrants to join the service industry may make the industrial environment less hostile for existing companies.

8.4 The enacted business environment: business strategies

8.4.1 Generic directions of business strategies

In order to carry out research analysis, five groups were assigned as prospectors, defenders, balancers, analysers and reactors. From the empirical results, most of the participating service organisations employ well-defined strategies guiding their business practice in East Asia. An analyser strategy was the most popular option selected by service organisations. Most organisations used multiple strategies such as balancers and analysers. On the whole, a defender strategy was the least favoured option for the corresponding executives.

Defender organisations attempted to devote their attention to the operational efficiency of their existing businesses while avoiding rapid adjustments in the organisational structure of their methods of operation. They tended to ignore developments outside established businesses and focused on well-defined customer groups with a full set of products or services.

Prospectors sought to initiate changes in their industrial sector and led innovation in the development of new products or services. Such organisations offered a broad range of products or services to the markets they served and grew mainly through product and market development as well as diversification.

Organisations following an analyser strategy tended to adapt rapidly to the changes developed by others. Some of their products or services and markets were stable while others were changing.

The essential feature of balancers is that they combined the characteristics of defender, prospector and analyser. There were three types of scenario with regard to their attitudes towards change: they seldom made a rapid adjustment to their existing businesses, but were able to adapt to changes created by others and meanwhile, they were capable of initiating changes within the industry.

Reactor organisations had the following attributes that weaken organisational strategic performance. First, their management or structure was not linked to the established strategy in an appropriate manner. Second, they adhered to existing structures or methods of operation, even though these were no longer relevant to environmental conditions. Third, they frequently perceived crucial changes occurring but were not always ready to respond. Finally, they made changes only when forced to do so by pressure.

8.4.2 Competitive strategies

There are four categories for assigning the competitive strategies of service organisations: low cost, differentiation, hybrid (or dual) and no-purpose. A differentiation strategy aims at achieving, even at considerable cost, a superior quality throughout the value chain and creating the image of a unique feature. The emphasis of a low-cost strategy is on lowering cost more than competitors wherever possible. A hybrid strategy means that a firm seeks to deploy more than one of the generic strategies and achieves cost leadership and differentiation simultaneously. If a firm fails to develop its strategy in at least one of the three directions, or is inconsistent in pursuing the generic strategies, and achieves no competitive advantage, the firm has no distinctive strategy and such firms are then defined as no-purpose strategy organisations.

In the empirical study, a differentiation-oriented competitive strategy was the major strategic selection, whereas a low-cost strategy was not a preferred choice within the industry. This was supported by the findings from each of the three countries. A higher proportion of Singaporean than Chinese and Malaysian organisations were keen on differentiating themselves from others.

8.4.3 Strategic positioning

Furthermore, from a strategic aspect, service organisations were categorised in terms of five competitive positions: high value premium (i.e. above the moderate level) price; high value moderate price; high value low (i.e. below the moderate level) price; low or moderate value low price; and uncompetitive value and price. The most frequent situation was the position with high customer added value, moderate or premium prices.

When comparing the three countries, the majority of the Singapore and Malaysia-based organisations pursued a combined advantage of high value and competitive (low and moderate) price. For China, the majority of the organisations were in the group of uncompetitive price and value.

To conclude, the strategic position of Singapore-based organisations appeared better than that of China-based organisations. Most of the high value premium organisations were from Singapore. The majority of the organisations following a strategy destined for ultimate failure with both uncompetitive price and value were found to come from China.

8.5 The outcomes of strategic performance

From the empirical work, overall, the oil and gas service organisations improved their strategic performance over the period examined, indicating a favourable situation for this industrial sector in East Asian countries like China, Singapore and Malaysia.

No significant differences in strategic performance were found for the groups between China and Malaysia or between China and Singapore, while the Malaysian group outperformed the Singaporean group significantly. The reasons that Malaysian organisations outperformed Singaporean organisations could be strong government support or easier access to oil clients as they are closer to oilfields than Singapore-based organisations.

The performance improvement was mainly attributed to the increase in total assets and business growth. However, their performance was offset by price competition, contractors' ability to influence customers' purchasing decisions, the speed of innovation and implementation of changes.

Since the marketplace is a congenial environment where organisations can perform well, the situation encourages service organisations, especially those international ones that take brave steps to gain entry into the marketplace. As the industry grows, more and more energy service entities such as those separate from indigenous or foreign-based oil organisations are becoming members of this big family to serve the oil and gas industry in the region.

8.6 The ESP alignments

The ESP (environment, strategy and performance) alignments here mean the relationships between the pairs of variables: business environment and business strategy, business environment and strategic performance, and business strategy and strategic performance.

8.6.1   Perceived business environment among strategic groups

In each of the three countries, different strategic options did not make any difference to the level of perceived uncertainty. Compared to reactors, defenders, prospectors and analysers, the degree of balancers' perceived environmental uncertainty was lower.

The results also indicate that organisations with strategies under the other four categories shared similar views on environmental uncertainty. In each of the three countries, research findings show that different strategic options did not make any difference to the level of perceived uncertainty.

8.6.2   Strategic performance by strategic groups

Among the various strategic directions followed by businesses, the levels of their strategic performance differ.

Balancer organisations outperformed all businesses under the other strategic dimensions, followed by analysers. Defender organisations were not associated with a higher level of strategic performance yet prospectors' strategic performance appeared to be better than defenders.

Organisations with a high level of strategic performance were not a reactive type. Rather, organisations in a low strategic performance context were inclined to embrace a reactor strategy. This suggests that a reactor strategic direction tended to obstruct organisational strategic performance.

In short, analysers or balancers yield a better strategic performance than reactors; differentiation or hybrid organisations outperform low-cost organisations; and a strategic position of low customer added value with a high price level is doomed to have a poor strategic performance.

General results were partially supported by the service organisations in the three countries. Regardless of strategic types, organisations in both China and Malaysia were likely to improve their strategic performance. However, there was not a significant linkage between business strategies and strategic performance in the above two countries.

In the research analysis, the question whether the level of strategic performance is related to the use of multiple strategies was examined.

This hypothesis can be applied in Singapore but not for China and Malaysia-based organisations. In Singapore, the higher level of strategic performance was associated with using multiple strategies; the lower level of strategic performance was associated with the use of non-multiple strategies. On the other hand, in China and Malaysia, service organisations could achieve sound strategic performance, no matter what types of business strategies were selected.

Since multiple strategies could yield oil and gas service organisations better strategic performance, the balancer and analyser strategies are concluded to be the most appropriate for managerial selection in an East Asian context.

8.6.3 Business environment versus strategic performance

There was a significant relationship between strategic performance and perceived environmental uncertainty or perceived hostility. Since the correlations are negative, it can be concluded that as the degree of perceived uncertainty decreases, there is a corresponding improvement in strategic performance. To support the above, findings derived from each of the three countries lead to the same conclusion.

8.7 Discussions of findings and proposition verifications

This section summarises the responses of the 98 organisations as a group in light of the 10 propositions. Ten propositions were examined in Chapters 4–6. The statistical significance levels in relation to the questionnaire responses were also calculated in the three analytical chapters.

The results that emerged show that some of the statistical findings were significant at the 0.01 or 0.05 levels or marginally significant at the 0.10 level, whereas other findings were not statistically significant. In this study, Propositions 5 and 6 were shown to be well grounded whereas Propositions 4, 9 and 10 were not. In addition, evidence was also not able to prove Proposition 8.

Proposition 1

The six environmental sectors – technology, regulation, economics, customers, suppliers and competitors – can be defined as the key task environmental sectors which oil and gas service executives perceive to be significant for the growth of their businesses in East Asia.

The research results support this contention. Firstly, it is demonstrated that each of the six task environmental factors has been important for oil and gas service companies' businesses in East Asia. Secondly, the economic, technological, customers' and competitors' factors have had a strong impact on the growth of their businesses in the region. In contrast, the impact of regulatory factors has been more likely to be moderate and the impact of suppliers has been more likely to be weak. In particular, the customers' factor has the strongest impact on businesses and it is thus considered as extremely important.

Proposition 2

For oil and gas service companies that operate in East Asian countries such as China, Singapore and Malaysia, the nature of the business environment will be uncertain.

The research results support this proposition. Overall, the business environment in which service organisations operate is slightly uncertain. Among the three countries, perceived uncertainty is supported by the Singapore groups. Perceived environmental uncertainty in China is the lowest and the business environment is perceived to be certain in that country. Generally, Malaysian organisations' executives have a neutral view on environmental uncertainty, indicating that the business environment in which they operate can be perceived as neither certain nor uncertain.

Proposition 3

Oil and gas service organisations' executives in East Asia perceive that the business environment in which they operate will be complex, dynamic and hostile.

This proposition is supported in part by the research. Significant statistical evidence proved that the business environment for the oil and gas service sector is complex and dynamic. The perceived complexity and dynamism are supported by the results from each of the three countries. It is also perceived that the business environment is pleasant for operating businesses in East Asia. However, in each of the three countries, there is always a split between organisations' executives: they were either pleased or displeased with the business environment in which they operated.

In particular, for Singapore and Malaysia, executives' perceptions tend to have a neutral view between pleasant and unpleasant towards environmental hostility.

Proposition 4

Perceived environmental uncertainty will be associated with the three environmental dimensions of perceived complexity, dynamism and hostility.

The research results give limited support to Proposition 4. It has been proved that perceived uncertainty is positively related to perceived dynamism and hostility. The results show that the higher the degree of perceived environmental dynamism and hostility, the higher the degree of perceived uncertainty. Moreover, the results do not support the initial assumption that there is a relationship between perceived complexity and perceived uncertainty.

Proposition 5

Perceived environmental uncertainty will be associated with the influences of the task environmental factors.

The results support this proposition. It is proved that perceived environmental uncertainty has a positive correlation with the number or diversity (heterogeneity) of environmental factors. For example, the results suggest that the more complicated the government regulations, legislation and policies in the region where a service organisation operates in East Asia, or the more different the supply conditions (e.g. price, quality, speed or service) provided by the organisation's suppliers, the higher the degree of perceived environmental uncertainty.

The unpredictability of the changes emerging in the six task environment sectors (i.e. regulatory, technological, economic, customers, competitors and suppliers) is correlated to perceived environmental uncertainty. If the level of unpredictability of certain environmental factors becomes higher, the degree of business environmental uncertainty perceived by service executives also tends to be higher.

The results also indicate that the greater the degree of the difficulties of resource (tangible and intangible) availability or resource deterrence (e.g. industry rivalry and relationships with clients), the higher the degree of perceived environmental uncertainty. For the associated environmental factors, if the level of perceived hostility becomes higher, the degree of perceived uncertainty also tends to be higher.

Proposition 6

Perceived environmental complexity, dynamism and hostility will be associated with the influences of the task environmental factors for the oil and gas service sector in East Asia.

The results support this proposition. With respect to the three perceived environmental dimensions, they can be examined by looking at their associations with various environmental factors. First, perceived complexity is associated with the three factors of economic, technological and customers' conditions. The more complicated the knowledge required for understanding the economic situation in the region, or the higher the levels of technology involved in the oil and gas service sector, or the more diversity in the needs and preferences of oil and gas clients, the higher the level of perceived environmental complexity.

Second, perceived dynamism is associated with only one factor in the economic sector: the higher the level of unpredictability of rig counts, the higher the degree of perceived dynamism. Perceived environmental dynamism has no relationship with the variables of the other task environmental sectors. This means that, from year to year, even though the changes in the regulatory, technological, customers', suppliers' and competitors' sectors cannot be forecast, each of these remaining task environmental sectors is not subject to a dynamic situation. Hence, for the oil and gas service organisations in East Asian countries like China, Singapore and Malaysia, perceived environmental dynamism is correlated with the unpredictability of the oil economic conditions.

Third, with reference to perceived hostility, it is associated with relevant customers' and competitors' conditions. The more distant the relationships with clients or the more turbulent the competition, the higher the degree of perceived hostility.

It can therefore be concluded that the economic, technological, competitors' and customers' conditions are more essential than regulatory and suppliers' influences in contributing to the nature of the three business environmental characteristics. This is consistent with the earlier results that the key environmental indicators are these four environmental sectors for service organisations in East Asia.

Proposition 7

For oil and gas service organisations operating in East Asian countries like China, Singapore and Malaysia, their business strategies will be different.

The results support this proposition. First, organisations can be split up in terms of five strategic groups: prospectors, defenders, balancers, analysers and reactors. Second, four competitive strategies of low cost, differentiation, hybrid (or dual) and no-purpose can be assigned for categorising service organisations. Third, organisations can be classified into five strategic competitive positions.

The majority of the participating organisations were balancers or analysers. The defender (rather than reactor) strategy was not an option preferred by senior executives. Similarly, the majority pursued differentiation-oriented generic strategies; in contrast, a low-cost competitive strategy was not an option preferred by senior management. There was not a significant majority in favour of assigning strategic competitive positions. This was contrary to the author's assumption that the majority of organisations would be keen on a strategic position of high value competitive price, whereas the minority fell into a category of uncompetitive price and value.

Proposition 8

For oil and gas service organisations operating in East Asia, managerial perceptions of the business environmental uncertainty will vary in association with the type of their strategic orientations.

There is insufficient evidence to support this proposition. First, none of the generic business strategies was found to be significantly related to perceived environmental uncertainty. Regardless of the generic strategic directions a service organisation follows, the category of perceived environmental certainty and uncertainty was not found to be associated with the strategic options adopted. Second, there is also not enough evidence to prove the hypothesis that the two variables of competitive strategies or strategic positions and the category of perceived uncertainty are in some way related. In each of the three countries, though different strategic options are conducted by organisations, it does not make any difference to the level of perceived uncertainty.

Conclusions drawn from the findings indicate that there is no significant relationship between strategic options and perceived environmental uncertainty. The current evidence fails to prove that the type of generic business strategies or strategic positions can provide any hints whether or not the perceived business environment could be uncertain.

Proposition 9

There will be relationships between the perceived business environmental dimensions and strategic performance for oil and gas service organisations operating in East Asia.

This proposition has gained limited support from the results. There is a significant relationship between strategic performance and perceived environmental uncertainty or perceived hostility. As the correlations are negative, it can be concluded that as the degree of perceived uncertainty or hostility decreases, there is a corresponding improvement in strategic performance.

Predominantly, results indicate a pattern that, when making a comparison among individual countries in an East Asian context, a higher level of strategic performance appears to be associated with a lower level of perceived environment uncertainty. When perceived uncertainty decreases at the same level, the corresponding improvement of strategic performance is different. China-based organisations improve in the same way as Singapore organisations do; by contrast, Malaysia-based organisations have a greater improvement in strategic performance than those based in China and Singapore.

What is more, insufficient evidence was provided to support the hypothesis that perceived environmental complexity or dynamism is associated with strategic performance. Hence, the research indicates that perceived complexity or dynamism is not significantly related to strategic performance.

The findings fail to provide evidence to support the assumption that the degree of perceived complexity or perceived dynamism is related to the level of strategic performance. Overall, the results support Proposition 9 that there is a relationship between two perceived business environmental dimensions and strategic performance for oil and gas service organisations operating in East Asia: the higher the perceived environmental uncertainty or hostility, the weaker the strategic performance.

Proposition 10

For oil and gas service organisations operating in East Asian countries like China, Singapore and Malaysia, strategic performance will differ in association with their business strategic orientations.

The proposition has gained limited support from the research results. In each country, results show that appropriate strategic options should essentially be relevant to a better performing business. It is observed that balancers and analysers outperformed the other categories of organisations. However, the analysis does not prove that reactors are associated with a relatively poor strategic performance, while the strategic performance of defenders and prospectors stands in the middle among the five categories of strategic orientations.

The assumption that hybrid organisations perform better than other organisations was disproved. It has been found that hybrid organisations yielded a higher level of strategic performance than low-cost and non-purpose strategic groups yet did not outperform differentiation organisations. The results also show that having a generic competitive strategy of low cost or differentiation or hybrid does not yield the organisation a higher level of strategic performance, while pursuing none of the three generic competitive strategies (no-purpose) produces a relatively poor strategic performance. In addition, having a differentiation strategy yields the service organisations in East Asia a higher strategic performance than the low-cost organisations.

Although organisations with high value competitive (i.e. low and moderate) price had a dual competitive advantage, they did not appear to perform better than organisations in all other categories of competitive positions. Organisations with uncompetitive value and price appeared to perform poorly in China, Singapore and Malaysia. However, the evidence is not sufficient to prove that organisations with various strategic competitive positions would have different strategic performance.

Overall, the statistics partially supported this book in that good or bad strategic performance is associated with certain strategic orientations adopted by service businesses in East Asia. Three themes should be stressed:

- Better strategic performance is associated with balancer or analyser strategies rather than with other types of strategy. As such, multiple strategy organisations (balancers and analysers) will outperform non-multiple strategy organisations (defenders, prospectors and reactors).
- Better strategic performance is associated with differentiation-oriented strategies rather than with a low-cost competitive strategy. Hence, a differentiation strategy is a more appropriate option than a low-cost strategy for service organisations operating in East Asia.
- Better strategic performance is associated with high customer added value premium prices rather than with other types of strategic competition position. Uncompetitive value and price organisations are guaranteed to have relatively poor strategic performance in each of the three selected countries.

8.8   Conclusions

In the introduction (Chapter 1) to this book, three objectives were proposed. To conclude, these have now been achieved successfully.

The first objective was to examine the dominant business environmental conditions which affect the oil and gas service industry in China and, to a lesser extent, in Singapore and Malaysia.

The book has depicted environmental uncertainty by evaluating environmental factors of the six task environmental sectors (i.e. customers, suppliers, competitors, regulatory, technology and economic) from three dimensions of complexity, dynamism and hostility. Propositions 1–6 presented above in this chapter provide detailed evidence to support this particular research objective.

The second objective was to empirically investigate strategies adopted by oil and gas service companies in response to the business environment(s) in which they operate.

Within the theoretical context, generic business strategies, competitive strategies employed and strategic positions achieved by oil and gas service organisations were investigated. Proposition 7 provided evidence in support of this research objective.

The final objective was to evaluate the reliability of the strategic theoretical frameworks based upon Western business practice applied in a non-Western business environment like East Asia. This objective has three foci: to evaluate the significance of environmental influences towards service organisations' long-term strategic success; to identify the effective business strategies which should be employed in an East Asia context; and to observe how strategic options can be selected in response to the business environment. Propositions 8–10 are used to evaluate the third objective.

On the whole, firms that consistently outperform over a period may have made a more accurate interpretation of the business environment. They can be more confident in coping with environmental influences. Hence, improved strategic performance is associated with the lower level of perceived environmental uncertainty.

Furthermore, service organisations with outstanding performance in the East Asian business environment pursued multiple business strategies.

In the light of the empirical evidence, it can be concluded that some strategic theoretical frameworks based upon Western business practice can be applied. Conversely, some others cannot be applied directly in an East Asian context unless they are developed according to the local situation.

In Singapore, the dominant players are foreign (mainly Western) organisations. The service sector in Singapore is mature and organisational practice is relatively standardised or westernised. The proposed strategic models based upon Western theories are therefore more applicable there than in China and Malaysia.

As introduced in Chapters 2 and 7, compared with Singapore, the history of the offshore oil and gas industrial development in China and Malaysia is shorter. In particular, the offshore service sector in China is fairly young as the Chinese government has been conducting a process to reform the oil and gas industry since the 1990s. In fact, indigenous companies (which usually have a government background) play a vital role within the service sector in China and Malaysia. Hence, these two markets are non-Western standardised and one outcome is that the strategic applications were not working very well in China and Malaysia when this study was being carried out.

Consequently, in the context of China and Malaysia, business environmental issues are crucial. It can be concluded that when initiating businesses in an immature or a non-Western standardised business environment in East Asia, senior executives should devote attention to significant environmental indicators and be confident in coping with the business environment. For the long term, when the majority of organisations practise Western strategic management theories, the application of appropriate business strategies should be essential for outstanding strategic performance. This is the case from what we can see in Singapore.

8.9   Contributions of the study

Overall, the merit of this study is that it should prove valuable to both business strategists (or those responsible for formulating strategy for business) and to researchers in academic, industrial and government spheres. Oil service organisations in East Asia in particular will find that the conclusions can assist them in devising profitable strategies for their businesses in the region. Moreover, having gathered empirical evidence from the interview and questionnaire survey of firms that operate in that area, this research will have made a significant addition to the existing strategic management literature. This claim must, however, be substantiated by analysis of both the theoretical and methodological contribution together with a brief account of the practical implications for the industry in East Asia.

8.9.1   Theoretical contribution

This study has contributed new data to the theory of strategic management. It has looked at existing literature based on Western strategic models, frameworks and theories and examined them in Eastern contexts. Having completed the study, a better understanding of whether Western strategic theories can be adapted for successful application in East Asia was obtained. The following demonstrates how Western models could be adjusted for Eastern application.

First, in the strategic management literature pertaining to the business environment, most writings have addressed issues such as industry driving forces, key success factors or major environmental indicators. Through this study, these themes can be evaluated quantitatively by using available models developed in this book. Furthermore, the applications of relevant strategic management theories have been examined and evaluated in the context of the energy service industry in East Asia. This achievement enables the author to fill in the existing theoretical gap for similar fields of study.

Second, the research results have observed that employing appropriate business strategies as defined in Western theory can yield organisations a good strategic performance. Strategic theories have been developed further to the extent that organisations with multiple business strategies outperform those without multiple business strategies in the context of the oil and gas service sector. Thus, the theory development and generated theoretical approach employed in this study are new contributions to knowledge.

8.9.2 Methodological contribution

The methodological contribution derived from this study is also important. It was assumed that conducting empirical research into senior executives operating in the oil and gas service industry in China, Singapore and Malaysia would be a huge and very difficult task. During the pilot study phase, it was found that senior managers might be puzzled about the purpose of the researcher's visits for interviewing and keenly aware of information confidentiality. In addition, the political sensitivities of senior managers in China, especially those in state-owned enterprises, may also contribute to the difficulties of conducting primary research in the country. Other crucial difficulties continually surfaced during the study. There was a lack of authoritative information on the East Asian oil and gas service sector and on the service organisations in China and Malaysia. Due to the continuous restructuring of the oil and gas industry in China, the total sample population was changing and therefore unknown. These difficulties are compounded by the fact that some Chinese senior managers were not familiar with using English and then the researcher had to communicate with respondents in Chinese when necessary.

It is considered challenging for any researcher channelled into Western thinking to conduct empirical work in East Asia. This study developed a methodology for data collection and analysis in China,

Singapore and Malaysia. The combined research methodology designed for this study contains both qualitative and quantitative attributes. The methodology could be applied to other countries and industries in the region. In particular, the techniques employed in this study for cross-national comparisons can also be applied by researchers who wish to conduct similar work.

In addition, there are two major aspects for the use of the survey questionnaire. Firstly, the researcher has generated a contingency approach to gaining access to the potential senior executive respondents. A range of industry contacts in China, Singapore and Malaysia was established in order to reach the potential respondents. Based upon the rapport established with the industry informants, the researcher then took the opportunity to ask for their knowledge or further information about the oil and gas service companies in the region of study. Eventually, a mailing list was prepared for the questionnaire survey. Five hundred questionnaires were distributed to the senior managers via post, email and fax at the first phase. At the second phase, follow-up phone calls, mails and emails were used in an attempt to achieve the proposed response target. A response rate of 21.6 per cent (China, 18 per cent; Singapore, 23 per cent; Malaysia, 14 per cent) not only enabled the researcher to gain enough data for statistical analysis, but also prove the success of the use of this contingency approach. It was observed that indigenous managers in China and Malaysia are more likely to respond to someone with whom they are familiar or someone they know in advance or someone introduced by their existing middle contacts. In the case of China, the middle person is also called a *guanxi* (personal relationship). This sort of obstacle could be the reason behind the relatively lower response rates from China and Malaysia. In these two countries, more efforts such as direct phone calls should be made in order to get a response from senior executives.

Secondly, as the researcher intended to reveal a wide picture of the energy service industry in the selected countries, the survey questionnaire is believed to be more informative and comprehensive than a purely qualitative exercise such as a case study. In the survey carried out for this study, there was a common focus reflected in the ten propositions related to the nature of the business environment, strategies and performance. The purpose of these propositions was to ensure that all responses could generate a common approach towards the issues under scrutiny.

8.9.3 Practical implications for the industry sector in East Asia

The practical implication of this study is the contribution of knowledge to the industry. As the firms under investigation operate in East Asia, the findings from this research can enrich knowledge of the oil and gas and wider energy service industries and the service sector in particular. Conclusions drawn from this study can be useful in practice for industrial experts and management to tackle particular issues when formulating strategic options.

The approach used for the assessment of environmental uncertainty in this study helps senior executives to identify the environmental situations. Therefore, the most appropriate methods such as scenario planning or forecasting for formulation of strategies can be taken into account after making an accurate appraisal of the business environment.

Appendix: Survey Questionnaire

The business environment and strategies for the energy service sector in East Asia (survey questions)

SECTION (I): BACKGROUND

Your Organisation
Tick √ only one as appropriate

1. Define the legal status of the organisation for which you are responsible.

☐ Independent company
☐ Division/Subsidiary Company

☐ Operating/Business Unit
☐ Other (Please specify) ....

Organisation location
Tick √ only one as appropriate

2. In which country is the organisation as defined above situated?

☐ China   ☐ Singapore

☐ Malaysia   ☐ Other ....

Size
Tick √ only one as appropriate

3. How many full time (or equivalent) people are employed in the above organisation?

☐ 1–49   ☐ 50–199   ☐ 200–499   ☐ 500–2999   ☐ >3000

Age

4. How long has the above organisation been in business under its present form?

Please specify: .... years

Ownership
Tick √ only one as appropriate

5. Identify the type of ownership of your organisation.

☐ Wholly Domestic State Owned
☐ Wholly Domestic Private/Individual
☐ Wholly Foreign Owned

☐ Joint Venture
☐ Domestic Share Holding/Public Limited (Company)
☐ Other (Please specify) ....

Industry segment
Tick √ the box(es) as appropriate

6. Which energy industry does the above organisation service?

(Oil and gas sector)
☐ Upstream (E&P)
☐ Midstream (Transportation)
☐ Downstream (Refining/processing and marketing)
☐ Other oil and gas service companies

(Other energy sector)
☐ Hydropower
☐ Electricity
☐ Nuclear
☐ Other Energy (Please specify) ....

7. For which oil and gas business activities do you provide a service?

☐ Exploration   ☐ Appraisal   ☐ Development   ☐ Production   ☐ Other ....

Business in East Asia
Tick √ the box(es) as appropriate

8. In which East Asian countries does this organisation operate oil and gas service businesses at present?

☐ China   ☐ Singapore   ☐ Malaysia   ☐ Philippines   ☐ Indonesia
☐ Vietnam   ☐ Thailand   ☐ Japan   ☐ South Korea   ☐ Other ....

Title of person responding

9. Please specify:
....

251

SECTION (II): THE BUSINESS ENVIRONMENT

Part One   **Environmental Sectors**

The business environment consists of the external factors considered to be crucial to the growth of your business within the oil and gas industry. These factors can be grouped into the following six sectors:

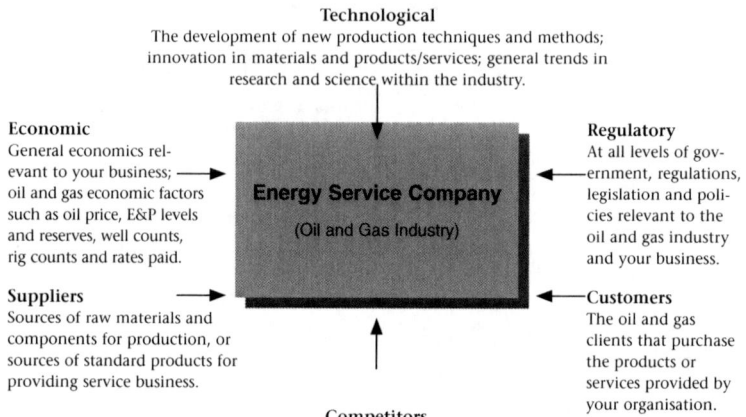

Technological
The development of new production techniques and methods; innovation in materials and products/services; general trends in research and science within the industry.

Economic
General economics relevant to your business; oil and gas economic factors such as oil price, E&P levels and reserves, well counts, rig counts and rates paid.

Energy Service Company
(Oil and Gas Industry)

Regulatory
At all levels of government, regulations, legislation and policies relevant to the oil and gas industry and your business.

Suppliers
Sources of raw materials and components for production, or sources of standard products for providing service business.

Customers
The oil and gas clients that purchase the products or services provided by your organisation.

Competitors
All competitors (existing firms, new entrants and firms providing substitute products/services); competitive tactics and actions among firms within the industry.

If a small change within an environmental sector brings about a great consequence to business, that sector is therefore considered **important**. However, it may be that a big change within that sector makes little difference to business and, on that basis, the sector may be considered to be of **no or little importance**. It is inevitable that each environmental sector, whatever its level of importance, will have a varying effect (therefore **impact**) on your business. Please circle the number that best indicates the above environmental sectors as they have affected your business over the **LAST 5 YEARS**:

1. In your opinion, how **IMPORTANT** are these sectors on the growth of your business?

		not important	slightly important	important	very important	most important
a)	Economic	1	2	3	4	5
b)	Technological	1	2	3	4	5
c)	Regulatory	1	2	3	4	5
d)	Customers	1	2	3	4	5
e)	Competitors	1	2	3	4	5
f)	Suppliers	1	2	3	4	5

2. For your experience, how does the **IMPACT** of these sectors affect the growth of your business?

	non-existent/ very weak	weak	moderate	strong	very strong
a) Economic	1	2	3	4	5
b) Technological	1	2	3	4	5
c) Regulatory	1	2	3	4	5
d) Customers	1	2	3	4	5
e) Competitors	1	2	3	4	5
f) Suppliers	1	2	3	4	5

Part Two   Nature of the Business Environment

For different industries, the nature of the business environment can mean different things. In each of the following statements, you may select <u>only one</u> by circling the number you consider most accurately reflects your organisational environment over the **LAST 5 YEARS**.

(1 and 7 = very/dramatically/strongly, 2 and 6 = exactly, 3 and 5 = tend to be, and 4 is neutral)

1. The business environment in which you operate is
   very static   1 2 3 4 5 6 7   *very dynamic*

2. The business environment in which you operate is
   very simple   1 2 3 4 5 6 7   *very complex*

3. The business environment in which you operate is
   very pleasant   1 2 3 4 5 6 7   *very unpleasant*

4. The knowledge required to understand the economic situation is
   very simple   1 2 3 4 5 6 7   *very sophisticated*

5. The market demand within the oil and gas industry which you serve is
   very big   1 2 3 4 5 6 7   *very small*

6. The market demand within the oil and gas industry which you serve is
   increasing dramatically   1 2 3 4 5 6 7   *decreasing dramatically*

7. The level of technology involved in the oil and gas service sector in which you operate is
   very low   1 2 3 4 5 6 7   *very high*

8. The products/services provided within the service sector in which you operate are
   very similar to each other 1 2 3 4 5 6 7   *very different from each other*

9. In the region where you operate, access to available technologies is
   very easy   1 2 3 4 5 6 7   *very difficult*

10. Government regulations, legislation and policies in the region where you operate are
    very simple   1 2 3 4 5 6 7   *very complicated*

11. The national government regulations and legislation in the region where you operate
    strongly benefit your business 1 2 3 4 5 6 7   *strongly limit your business*

12. The local level government policies (e.g. Customs or administrative bureaux) in the region where you operate

very positively influence your business   1 2 3 4 5 6 7   *very negatively influence your business*

13. The relationships of your organisation with government are

very close   1 2 3 4 5 6 7   *very distant*

14. Within the oil and gas industry, the number of customers whom you serve is

very small   1 2 3 4 5 6 7   *very large*

15. The needs and preferences of the oil and gas clients whom you serve are

very similar   1 2 3 4 5 6 7   *very different*

16. Your key customers switch to another competitor's products/services

with great difficulty   1 2 3 4 5 6 7   *very easily*

17. The relationships of your organisation with your key oil and gas clients are

very good   1 2 3 4 5 6 7   *very poor*

18. The number of suppliers to your organisation are

very small   1 2 3 4 5 6 7   *very large*

19. The supply conditions (e.g. price, quality, speed, or service) provided by your suppliers are

very similar   1 2 3 4 5 6 7   *very different*

20. Access to suppliers for obtaining available raw materials or standard goods and services is

very easy   1 2 3 4 5 6 7   *very difficult*

21. The relationships of your organisation with your key suppliers are

very supportive   1 2 3 4 5 6 7   *very unhelpful*

22. The number of firms within the service sector in which you operate is

very small (market is dominated by a few firms)   1 2 3 4 5 6 7   *very large (market is shared by numerous firms)*

23. The scope of firms within the service sector in which you operate is

very narrow (completely domestic companies only)   1 2 3 4 5 6 7   *very extensive (companies from all over the world)*

24. The entry barriers to the oil and gas service sector in which you operate are

very high   1 2 3 4 5 6 7   *very low*

25. The rivalry among the competitors within the service sector in which you operate is

very orderly   1 2 3 4 5 6 7   *very turbulent*

26. Competitive actions adopted by firms within the service sector in which you operate are

very reasonable   1 2 3 4 5 6 7   *very unreasonable*

27. The relationship of your organisation with your key competitors is

very collaborative   1 2 3 4 5 6 7   *very uncollaborative*

28. The business environment in which you operate is

very certain   1 2 3 4 5 6 7   *very uncertain*

The following questions pertain to the degree of predictability taking place in each of the same six environmental sectors as defined above. PREDICTABLE means the situation stays about the same from year to year. UNPREDICTABLE means things that cannot be forecast from year to year. How would you rate the following environmental factors on the degree of their predictability in the **LAST 5 YEARS**?

1	2	3	4	5	6	7
highly predictable	predictable	tend to be predictable	neutral	tend to be unpredictable	unpredictable	highly unpredictable

(Circle one)

29. Customers
    a) demand for existing products/services   1 2 3 4 5 6 7
    b) demand for new products/services   1 2 3 4 5 6 7
    c) demand for higher quality or more services   1 2 3 4 5 6 7
    d) preference for lower price   1 2 3 4 5 6 7

30. Competitors
    a) changes in their competitive price of products/services   1 2 3 4 5 6 7
    b) improvement in quality of products/services   1 2 3 4 5 6 7
    c) introduction of new products/services   1 2 3 4 5 6 7

31. Suppliers
    a) rising prices   1 2 3 4 5 6 7
    b) reduction in quality of goods and services   1 2 3 4 5 6 7
    c) introduction of new materials and components   1 2 3 4 5 6 7
       or standard products

32. Economic
    a) changes in oil and gas exploration and production   1 2 3 4 5 6 7
       in the region where you operate
    b) changes in well counts in the region where   1 2 3 4 5 6 7
       you operate
    c) changes in rig activities (rig counts) in the   1 2 3 4 5 6 7
       region where you operate

33. Technological
    a) technological changes in the service sector   1 2 3 4 5 6 7
       in which you operate
    b) rate of technological diffusion throughout the   1 2 3 4 5 6 7
       service sector in which you operate

34. Regulatory
    a) changes in national regulations and legislation   1 2 3 4 5 6 7
       in the region where you operate
    b) changes in local level government policies   1 2 3 4 5 6 7
       in the region where you operate

SECTION (III): STRATEGY

Part Three   **Strategic Orientation**

Strategy is defined as a pattern of organisational decisions for positioning a firm in the environment and guiding internal operations. From your knowledge, please assess each of the following statements by selecting the most applicable to your organisation over the **LAST 5 YEARS.**

1	2	3	4	5	6	7
strongly agree	agree	tend to agree	neutral	tend to disagree	disagree	strongly disagree

(Circle one)

1. We have clearly articulated our business strategy(ies).   1 2 3 4 5 6 7
2. We offer a broad range of products/services to the markets we serve.   1 2 3 4 5 6 7
3. We direct our product/services to well-defined customer groups.   1 2 3 4 5 6 7
4. We offer customers a full set of products/services.   1 2 3 4 5 6 7
5. Some of our products/services and markets are stable, others changing.   1 2 3 4 5 6 7
6. We seek to increase use of existing products/services in existing markets.   1 2 3 4 5 6 7
7. We seek to develop some products/services that are closely related to our existing products/services in existing markets.   1 2 3 4 5 6 7
8. We seek to develop new markets for our existing products/services.   1 2 3 4 5 6 7
9. We seek to introduce new products/services in our existing markets.   1 2 3 4 5 6 7
10. We seek to generate unrelated new products/services and markets.   1 2 3 4 5 6 7
11. We seek to create new products/services to make similar existing products/services obsolete.   1 2 3 4 5 6 7
12. We devote attention to the operational efficiency for our existing business.   1 2 3 4 5 6 7
13. We tend to ignore developments outside our established business.   1 2 3 4 5 6 7
14. We seldom make rapid adjustments in our organisational structure or methods of operation.   1 2 3 4 5 6 7
15. We have a viable strategy but our management or structure is not linked to it in an appropriate manner.   1 2 3 4 5 6 7
16. We adhere to our structure or methods of operation even though they are no longer relevant to environmental conditions.   1 2 3 4 5 6 7
17. Within a single business activity, we switch between emphasising factors such as quality, unique features, price or other products/services attributes.   1 2 3 4 5 6 7

18. We emphasise quality, price, unique features or other products/services attributes depending on the business activity concerned.   1 2 3 4 5 6 7

19. We frequently perceive crucial change occurring but are not always ready to respond.   1 2 3 4 5 6 7

20. We rapidly adapt to the change developed by others.   1 2 3 4 5 6 7

21. We usually create change in the industry.   1 2 3 4 5 6 7

22. We make changes only when forced to do so by pressures.   1 2 3 4 5 6 7

23. We seek to provide the highest possible quality of business or products/services.   1 2 3 4 5 6 7

24. We tend to emphasise a unique feature of business or products/services.   1 2 3 4 5 6 7

25. We seek to achieve the lowest possible cost of business or products/services.   1 2 3 4 5 6 7

26. We compete based on the highest quality of business or products/services.   1 2 3 4 5 6 7

27. We differentiate our business or products/services from those of our competitors.   1 2 3 4 5 6 7

28. We compete based on the lowest cost of business or products/services.   1 2 3 4 5 6 7

29. Within the oil and gas service industry, each time we compete with other companies in a marketplace in East Asia,

(tick √ <u>the boxes</u> as appropriate)

a) price charged by our products/services is

☐ Very low ☐ Below moderate ☐ Moderate ☐ Above moderate ☐ Very high

b) products/services quality valued by customers is

☐ Very low ☐ Below moderate ☐ Moderate ☐ Above moderate ☐ Very high

c) reliability of (products/services) technology is

☐ Very low ☐ Below moderate ☐ Moderate ☐ Above moderate ☐ Very high

d) safety performance of products/services is

☐ Very low ☐ Below moderate ☐ Moderate ☐ Above moderate ☐ Very high

e) speed of responding to clients' requirements is

☐ Very low ☐ Below moderate ☐ Moderate ☐ Above moderate ☐ Very high

f) price that customers are willing to pay is

☐ Much lower than what we offer   ☐ Below what we offer   ☐ The same as what we offer   ☐ Above what we offer   ☐ Much higher than what we offer

Part Four   Strategic Performance

This part is concerned with how your organisation's performance **changed** in the **PAST 5 YEARS**. In each of the following questions, to your knowledge, you may select **only one** by circling the number that best estimates your organisation's situation.

1	2	3	4	5	6	7
much less	less	slightly less	no change	slightly more	more	much more

1. Total assets (current and fixed assets)   1 2 3 4 5 6 7
2. Return on total assets (net profits after taxes   1 2 3 4 5 6 7
   to total assets)
3. Annual sales/production revenue   1 2 3 4 5 6 7
4. Annual net profit margin (net profits after taxes to   1 2 3 4 5 6 7
   sales/production revenue)
5. Total asset turnover (sales/production revenue to total assets)   1 2 3 4 5 6 7

1	2	3	4	5	6	7
much worse	worse	slightly worse	no change	slightly better	better	much better

6. Products/services cost improvements   1 2 3 4 5 6 7
7. Price competitiveness   1 2 3 4 5 6 7
8. Products/services reliability (cost involved, ease of use,   1 2 3 4 5 6 7
   speed of delivery, technical support, service availability, etc.)
9. Products/services quality   1 2 3 4 5 6 7
10. Relationship with key customers   1 2 3 4 5 6 7
11. Capability of influencing customers' purchase decisions   1 2 3 4 5 6 7
12. Optimism of obtaining/resuming contracts   1 2 3 4 5 6 7
13. Key employment stability   1 2 3 4 5 6 7
14. Personnel calibre at all levels   1 2 3 4 5 6 7
15. Operational efficiency (do things right first time)   1 2 3 4 5 6 7
16. Speed of innovation and implementation of change   1 2 3 4 5 6 7
17. Overall organisation effectiveness (do right things   1 2 3 4 5 6 7
    and eliminate non-positive impact activities)
18. Value added   1 2 3 4 5 6 7
19. General organisation image   1 2 3 4 5 6 7
20. Confidence to achieve growth   1 2 3 4 5 6 7

Bibliography

Books

Aldrich, H., 1979. *Organisations and Environments*. Prentice Hall, Englewood Cliffs, NJ.

Aldrich, H.E. and Mindlin, S., 1978. Uncertainty and Dependence: Two Perspectives on Environment. In: Kerpit, L. (ed.), *Organisation and Environment*, pp. 149–70. Sage, Beverly Hills, Calif.

Ansoff, H. I., 1965. *Corporate Strategy*. McGraw-Hill, New York.

—— 1979. *Strategic Management*. Macmillan, London.

Boyle, J. S., 1994. Styles of Ethnography. In: Morse, J. M. (ed.), *Critical Issues in Qualitative Research Methods*. Sage, London.

Brooks, I. and Weatherston, J., 1997. *The Business Environment: Challenges and Changes*. Prentice Hall, London.

Bryman, A., 1988. *Quantity and Quality in Social Research*. Unwin Hyman, London, p. 88.

Burrell, G. and Morgan, G., 1979. *Sociological Paradigms and Organisational Analysis*. Heinemann, London, pp. 6–7.

Carmines, E. and Zeller, R., 1979. *Reliability and Validity Assessment*. Sage, Beverly Hills, Calif.

Chandler, A. D., 1962. *Strategy and Structure: Chapters in the History of the Industrial Enterprises*. MIT Press, Cambridge, Mass., USA.

Chen, M., 1995. *Asian Management System: Chinese, Japanese and Korean Styles of Business*. International Thomson Business Press, London.

Christopher, M., 1998. *Logistics and Supply Chain Management*, 2nd edn. Financial Times Pitman Publishing, London.

Clausewitz, C., 1812. *Vom Kriege*, translated and edited by Hans W. Gatzkein in 1942. The Military Service Publishing Company, Harrisburg, Pa.

Colin, S. G., 2002. *Strategy for Chaos: Revolutions in Military Affairs and the Evidence of History*. Frank Cass, London.

Cooper, H., 1989. *Integrating Research: a Guide for Literature Review*, 2nd edn. Sage, Beverley Hills.

Crabtree, B. F. and Miller, W. L., 1992. *Doing Qualitative Research: Research Methods for Primary Care*, Vol. 3. Sage, London.

Cresswell, J. W., 1994. *Research Design: Qualitative and Quantitative Approaches*. Sage, Thousand Oaks, Calif.

Dey, I., 1993. *Qualitative Data Analysis: a User-Friendly Guide for Social Scientists*. Routledge, London.

Dill, W. R., 1962. The Impact of Environment on Organisational Development. In: Mailick S. and Edward, H. van Ness (eds), *Concepts and Issues in Administrative Behaviour*, pp. 94–109. Prentice Hall, Englewood Cliffs.

Drucker, P., 1970. *Technology, Management and Society: Essays.* Heinemann, London.
—— 1974. *Management: Task, Responsibilities, Practices.* Harper & Row, London.
—— 1980. *Managing in Turbulent Times.* Heinemann, London.
Everitt, B., 1986. *Cluster Analysis,* 2nd edn. Gower, New York.
Fahey, L. and Narayanan, V. K., 1986. *Macraoenvironmental Analysis for Strategic Management.* West, St. Paul, Minn.
Faulkner D. and Bowman, C., 1995. *The Essence of Competitive Strategy.* Prentice Hall, New York.
Field, A., 2000. *Discovering Statistics: using SPSS for Windows.* Sage, London.
Fox, D. J., 1969. *The Research Process in Education.* Holt Rinehart and Winston, New York.
Gill, J. and Johnson, P., 1991. *Research Methods for Managers.* Chapman, London.
Glaser, B.G. and Strauss, A.L., 1967. *The Discovery of Grounded Theory: Strategies for Qualitative Research.* Aldine De Gruyter, New York.
Green, S., Salkind, N. and Akey, T., 2000. *Using SPSS for Windows: Analysing and Understanding Data,* 2nd edn. Prentice Hall, New Jersey.
Gummesson, E., 1991. *Qualitative Methods in Management Research,* rev. edn. Sage, London.
Hofer, C. W. and Schendel, D., 1978. *Strategy Formulation: Analytical Concepts.* The West Series in Business Policy and Planning. West Pub. Co., St. Paul.
Hofstede, G. H., 1994. *Cultures and Organisations: Software of the Mind, Intercultural Cooperation and its Importance for Survival.* Fontana, London.
—— 2001. *Culture's Consequences: Comparing Values, Behaviour, Institutions and Organisations across Nations,* 2nd edn. Sage, Beverly Hills, Calif.
Hollenson, S., 1998. *Global Marketing: a Market-Responsive Approach,* 7th edn. Prentice Hall, London.
Johnson, G. and Scholes, K., 1999. *Exploring Corporate Strategy: Text and Cases,* 5th edn. Prentice Hall, London.
Kerr, A. W., Hall, H. K. and Kozub, S. A., 2002. *Doing Statistics with SPSS.* Sage, London.
Kim, J. and Mueller, C., 2000. Factor Analysis: Statistical Methods and Practical Issues. In: Lewis-Beck, Michael S. (ed.), *Factor Analysis and Related Techniques.* Sage/Toppan Publishing, London.
Klecka, W. R., Norman, N. H. and Hull, C. H., 1975. *SPSS (Statistical Package for the Social Sciences) Primer.* McGraw-Hill Book Company, New York.
Lawrence, P. R. and Dyer, D., 1983. *Renewing American Industry: Organising for Efficiency and Innovation.* Free Press, New York.
Levitt, T., 1969. *The Marketing Mode: Pathways to Corporate Growth.* McGraw-Hill, Maidenhead and New York.
Lewis-Beck, M., 1994. *Factor Analysis and Related Techniques.* International Handbooks of Quantitative Applications in the Social Sciences, Vol. 5. Sage, London.
Lu, G., 1999. *The Small and Medium Size Enterprises Research.* Shanghai Economics and Finance University Publishing, China.
Lynch, R., 2000. *Corporate Strategy.* Financial Times, Prentice Hall, London.

McDaniel, C. and Gate, R., 1993. *Contemporary Marketing Research*, 2nd edn. West Publishing Company, New York, p. 88.

McNamee, P. B., 1992. *Strategic Management: a PC-based Approach*. Butterworth-Heinemann Ltd, Oxford.

Mansfield, R., 1990. Conceptualising and Managing the Organisational Environment. In: Wilson, D.C. and Rosenfield, R.H. (eds), *Managing Organisation: Texts, Readings and Cases*. McGraw-Hill, London.

Markillie, P., 1999. *Asia Finds its Stride, the World in 2000*. The Economist Newspaper Limited, London, p. 81.

Miles, R.E. and Snow, C. C., 1978. *Organisational Strategy, Structure, and Process*. McGraw-Hill, New York.

Mintzberg, H., 1979. *The Structuring of Organisations: a Synthesis of the Research*. Prentice Hall, Englewood Cliffs, NJ, pp. 268–9.

Mintzberg, H., Ahlstrand, B. and Lampel, J., 1998. *Strategy Safari: a Guided Tour through the Wilds of Strategic Management*. Prentice Hall, London.

Nash, M., 1983. *Managing Organisational Performance*. Jossey-Bass, San Francisco.

Nunnally, J., 1978. *Psychometric Methods*. McGraw-Hill, New York.

Ogden, E. H., 1993. *Completing Your Doctoral Dissertation or Master's Thesis in Two Semesters or Less*. Technomic Publishing Co., Inc., Lancaster, UK.

Ohmae, K., 1982. *The Mind of the Strategist: the Art of Japanese Business*. McGraw-Hill, New York.

Oppenheim, A. N., 1992. *Questionnaire Design, Interviewing and Attitude Measurement*, 2nd edn. Pinter Publishers, London.

Paid, K. W., 1995. *Gas and Oil in Northeast Asia: Policies, Projects and Prospects*. Royal Institute of International Affairs, London.

Paludan, A., 1998. *Chronicle of the Chinese Emperors: the Reign-by-Reign Record of the Rulers of Imperial China*. Thames and Hudson Ltd, London.

Pearce II, J. and Smith, G., 1997. Note on the Oil and Gas Exploration and Production Industry. In: Pearce, J. A. and Robinson, R. B., *Strategic Management: Formulation, Implementation and Control*. IRWIN, London.

Pearce, J. A. and Robinson, R. B., 1997. *Strategic Management: Formulation, Implementation and Control*. IRWIN, London.

Pfeffer, J. and Salancik, G. R., 1978. *The External Control of Organisations*. Harper and Row Publishers, New York.

Porter, M. E., 1980a. *Competitive Advantage: Creating and Sustaining Superior Performance*. Collier Macmillan, London.

—— 1980b. *Competitive Strategy: Techniques for Analysing Industries and Competitors*. The Free Press, New York.

—— 1985. *Competitive Strategy: Techniques for Analysing Industries and Competitors*. The Free Press, New York.

Rees, W. D. and Porter, C., 2001. *Skills of Management*, 5th edn. Thomson Learning, London.

Robert, M., 2000. *The Power of Strategic Thinking: Lock In Markets, Lock Out Competitors*. McGraw-Hill, New York.

Robson, C., 1993. *Real World Research*. Blackwell, Oxford.

Saunders, M., Lewis, P. and Thornhill, A., 1997. *Research Methods for Business Students*. Financial Times, Pitman Publishing, London.

Scott, W. R., 1981. *Organisations: Rational, Natural, and Open Systems*. Prentice Hall, Englewood Cliffs, NJ.

Sondhi, R. 1999. *Total Strategy*. Airworthy Publications International Limited, Bury, Lancashire.

Strauss, A. and Corbin, J., 1990. *Basics of Qualitative Research: Grounded Theory Procedures and Techniques*. Sage, Beverly Hills, Calif.

Sun Zi, around 476 BC. *The Art of War* (in ancient Chinese). Hunan Press, Hunan.

Thompson, J. D., 1967. *Organisations in Action*. McGraw-Hill Book Company, New York.

Warner, M., 2000. *Changing Workplace Relations in the Chinese Economy*. Palgrave Macmillan, London.

Weick, K., 1979. *The Social Psychology of Organising*, 2nd edn, Addison-Wesley, Reading, Mass.

Wilson, D.C., 1992. *A Strategy of Change*. Routledge, London.

Yin, R.K., 1994. *Case Study Research: Design and Methods*. Applied Social Research Methods Series, Vol. 5, 2nd edn. Sage, London.

Zikmund, W. G., 2000. *Business Research Methods*, 6th edn. The Dryden Press, Harcourt College Publishers, Chicago.

Journals and periodicals

Abraham, K. S., 1999. Upstream Climate Brightening in Southeast Asia: a World of Oil. *World Oil*, December, p. 29.

Badri, M. A., Davis, D. and Davis D., 2000. Operations Strategy, Environmental Uncertainty and Performance: a Path Analytical Model of Industries in Developing Countries. *Journal of Operations Management*, 28 (2), pp. 155–73.

Barry, H., 1977. Strategy and the 'Business Portfolio'. *Long Rang Planning*, 10(1), pp. 9–15.

Bourgeois, L. J. 1980. Strategy and Environment: a Conceptual Integration. *The Academy of Management*, January, 5(1), pp. 25–39.

Boyd, W. B. and Westfall, R., 1970. Interviewer Bias Once More Revisited. *Journal of Marketing Research*, 7, pp. 249–50.

Campbell-Hunt, C., 2000. What Have We Learned about Generic Competitive Strategy? A Meta-Analysis. *Strategic Management Journal*, 21 (2), pp. 127–54.

Carter, N. M., Stearns, T.M, Reynolds, P.D. and Miller, B.A., 1994. New Venture Strategies: Theory Development with an Empirical Base. *Strategic Management Journal*, 15, pp. 21–41.

Chan, R. Y. and Wong, Y.H., 1999. Bank Generic Strategies: Does Porter's Theory Apply in an International Banking Centre? *International Business Review*, 8, pp. 561–90.

Chandler, G.N. and Hanks, S.H., 1993. Measuring the Performance of Emerging Businesses: a Validity Study. *Journal of Business Venturing*, 8, pp. 391–408.

Chen, G., 2000. The Development and Reform of the Chinese Oil and Gas Industry in the 21st Century. *International Petroleum Economics*, 8(1), p. 5.

Clarke, T. and Du, Y., 1998. Corporate Governance in China, Explosive Growth and New Patterns of Ownership. *Long Range Planning*, 31(2), pp. 239–51.

Croteau, A.M. and Bergeron, F., 2001. An Information Technology Trilogy: Business Strategy, Technological Development and Organisational Performance. *The Journal of Strategic Information Systems*, 10(2), pp. 77–99.

Daft, R. L., Sornumen, J. and Parks, D., 1988. Chief Executive Scanning, Environmental Characteristics and Company Performance: an Empirical Study. *Strategic Management Journal*, 9, pp. 123–39.

Dess, G. G. and Beard, D. W., 1984. Dimensions of Organisational Task Environments. *Administrative Science Quarterly*, 29, pp. 52–73.

Dess, G. G. and Robinson, R.B., 1984. Measuring Organisational Performance in the Absence of Objective Measures: the Case of the Privately-Held Firm and Conglomerate Business Unit. *Strategic Management Journal*, 5, pp. 265–73.

Dill, W. R., 1958. Environment as an Influence on Managerial Autonomy. *Administrative Science Quarterly*, 2, pp. 409–43.

Downey, H. K., Hellriegel, D. and Slocum, J. W. Jr, 1975. Environmental Uncertainty: the Construct and its Application. *Administrative Science Quarterly*, 20, pp. 613–29.

Duncan, R. B., 1972. Characteristics of Organisational Environments and Perceived Environmental Uncertainty. *Administrative Science Quarterly*, 17, pp. 313–27.

Elenkov D. S., 1997. Strategic Uncertainty and Environmental Scanning: the Case for Institutional Influences on Scanning Behaviour. *Strategic Management Journal*, 18(4), pp. 287–302.

Emery, F. E. and Trist, E. L., 1965. The Causal Texture of Organisational Environments. *Human Relations*, 18, pp. 21–31.

Feurer, R., Chaharbaghi, K. and Wargin, J., 1995. Analysis of Strategy Formulation and Implementation at Hewlett-Packard. *Management Decision*, 33, pp. 4–16.

Gemba, K. and Kodama, F., 2001. Diversification Dynamics of the Japanese Industry. *Research Policy*, 30(8), pp. 1165–84.

Hall, W.K., 1978. SBUs: Hot, New Topic in the Management of Diversification. *Business Horizons*, 21(1), pp. 17–25.

Hambrick, D. C., 1982. Environmental Scanning and Organisational Strategy. *Strategic Management Journal*, 3, pp. 159–74.

Hegarty, W. H. and Tihanyi, L., 1999. Surviving the Transition: Central European Bank Executives' View of Environmental Changes. *Journal of World Business*, 34 (4), pp. 409–22.

Helms, M. M., Dibrell, C. and Wright, P., 1997. Competitive Strategies and Business Performance: Evidence from the Adhesives and Sealant Industry. *Management Decision*, 35(9), pp. 689–703.

Hofer, C. W., 1975. Toward a Contingency Theory of Business Strategy. *Academy of Management Journal*, 18(4), pp. 784–810.

—— 1980. Turnaround Strategies. *Journal of Business Strategy*, 1(1) pp. 19–31.

Hotha, S. and Vadlamani, B. L., 1995. Assessing Generic Strategies: an Empirical Investigation of Two Competing Typologies in Discrete Manufacturing Industries. *Strategic Management Journal*, 16, pp. 75–83.

Hrebiniak, L. G. and Snow, C. C., 1980. Industry Differences in Environmental Uncertainty and Organisational Characteristics Related to Uncertainty. *Academy of Management Journal*, 23, pp. 750–9.

Igel, B. and Islam, N., 2001. Strategies for Service and Market Development of Entrepreneurial Software Designing Firms. *Technovation*, 21(3), pp. 157–66.

Ireland, R. D., Hitt, M. A., Bettis, R. A. and Porras, D., 1987. Strategy Formulation Processes: Differences in Perceptions of Strength and Weaknesses Indicators and Environmental Uncertainty by Managerial Level. *Strategic Management Journal*, 8, pp. 469–85.

Johnson, F., 2002. Napoleon and Hitler Would Have Been Hopeless at Business; So Why Is Business Obsessed with War? *The Spectator*, 9 March.

Koberg, C. S., 1987. Resource Scarcity, Environmental Uncertainty, and Adaptive Organisational Behaviour. *Academy of Management Journal*, 30, pp. 798–807.

Kotha, S. and Nair, A., 1995. Strategy and Environment as Determinants of Performance: Evidence from the Japanese Machine Tool Industry. *Strategic Management Journal*, 16, pp. 496–518.

Lasserre, P., 1995. Corporate Strategies for the Asia Pacific Region. *Long Range Planning*, 28(1), pp. 13–30.

Lawrence, P. and Lorsch, J. W., 1967. Differentiation and Integration in Complex Organisations. *Administrative Science Quarterly*, 12, pp. 1–47.

Lee, S.F., Roberts, P., Lau, W.S. and Bhattacharyya, S.K., 1998. Sun Tzu's The Art of War as Business and Management Strategies for World Class Excellence Evaluation under QFD Methodology. *Business Process Management Journal*, 4(2), pp. 96–113.

Li, J., 2000. The Competitive Strategy of China's Township Enterprises: Understanding the Sources for Survival and Success. *Business Process Management Journal*, 7(4), pp. 340–8.

Li, J., Qian, G., Lam, K. and Wang, D. 2000. Breaking into China: Strategic Considerations for Multinational Corporations. *Long Range Planning*, 33, pp. 673–87.

Low, S.P., 2001. Chinese Business Principles from the Eastern Zhou Dynasty (770–221 BC): Are They Still Relevant Today? *Marketing Intelligence and Planning*, MCB University Press, 19(3), pp. 200–7.

Low, S.P. and Lee, S.K., 1997. Managerial Grid and Zhuge Liang's Art of Management; Integration for Effective Project Management. *Management Decision*, 35(5), pp. 382–91.

Luo, Y. and Park, S. H., 2001. Strategic Alignment and Performance of Market-Seeking MNCs in China. *Strategic Management Journal*, 22, pp. 141–55.

Luo, Y.D. and Tan, J.J., 1998. A Comparison of Multinational and Domestic Firms in an Emerging Market: a Strategic Choice Perspective. *Journal of International Management*, 4(1), pp. 21–40.

McGee, J. and Segal-Horn, S., 1990. Strategic Space and Industry Dynamics. *Journal of Marketing Management*, 6(3).

Miller, D., 1992. The Generic Strategy Trap. *The Journal of Business Strategy*, January/February 1992.

Miller, K. D., 1993. Industry and Country Effects on Managers' Perception of Environmental Uncertainties. *Journal of International Business Studies*, 24, pp. 693–714.

Morse, J. M., 1991. Approaches to Qualitative–Quantitative Methodological Triangulation. *Nursing Research*, 40, pp. 120–3.

Parnell, J. A., Lester, D. L. and Menefee, M. L., 2000. Strategy as a Response to Organisational Uncertainty: an Alternative Perspective on the Strategy–Performance Relationship. *Management Decision*, 38(8), pp. 520–30.

Peng, W., Lu, Y., Shenkar, O. and Wang, Y.L., 2001. Treasures in the China House: a Review of Management and Organisational Research on Greater China. *Journal of Business Research*, 52, pp. 97–110.

Proff, H., 2000. Hybrid Strategies as a Strategic Challenge – the Case of the German Automotive Industry. *The International Journal of Management Science*, 28, pp. 541–53.

Ramanujam, V. and Venkatraman, N., 1987. Planning System Characteristics and Planning Effectiveness. *Strategic Management Journal*, 8, pp. 453–68.

Scott, C., 1961. Research on Mail Surveys. *Journal of the Royal Statistical Society*, 24(124), pp. 143–205.

Simerly, R. L. and Li, M., 2000. Environmental Dynamism, Capital Structure and Performance: a Theoretical Integration and an Empirical Test. *Strategic Management Journal*, 21, pp. 38–41.

Simons, R., 1987. Accounting Control Systems and Business Strategy: an Empirical Analysis. *Accounting Organisations and Society*, 12(4), pp. 357–60.

Slater, S.F. and Olson, E.M., 2000. Strategy Type and Performance: the Influence of Sales Force Management. *Strategic Management Journal*, 21, pp. 812–29.

Smircich, L. and Stubbart, C., 1985. Strategic Management in an Enacted Environment. *Academy of Management Review*, 10(4), pp. 724–36.

Snyder, R. E., 1999. What is Happening in Drilling. *World Oil*, December, p. 23.

Stern, P.N., 1980. Grounded Theory Methodology: its Uses and Process. *Image*, 12, pp. 20–3.

Sutcliffe, K. M. and Huber, G. P., 1998. Firm and Industry as Determinants of Executive Perceptions of the Environment. *Strategic Management Journal*, 19, pp. 793–801.

Sutcliffe, K. M. and Zaheer, A., 1998. Uncertainty in the Transaction Environment: an Empirical Test. *Strategic Management Journal*, 19, pp. 1–23.

Tan, J. J. and Litschert, R. J., 1994. Environment–Strategy Relationship and its Performance Implications: an Empirical Study of the Chinese Electronics Industry. *Strategic Management Journal*, 15, pp. 1–20.

Tan, Y., 2001. Employing Strategies to Adapt to the Business Environment: an Empirical Pilot Study in China, the 2001 NEBAA (New England Business Administration Association) International Conference, Scotland, May.

Tan, Y., Gourlay, D. and Capsey, M., 2001. Managerial Perceptions of the Business Environment: an Empirical Pilot Test in China. *American Society of Business and Behavioural Sciences Conference, London School of Economics, August.*

Tung, R. L., 1979. Dimensions of Organisational Environments: an Exploratory Study of their Impact on Organisation Structure. *Academy of Management Journal*, 22(4), pp. 672–93.

Venkatraman, N. and Presott, J.E., 1990. Environment–Strategy Coalignment: an Empirical Test of its Performance Implications. *Strategic Management Journal*, 11, pp. 1–23.

Venkatraman, N. and Ramanujam, V., 1986. Measurement of Business Performance in Strategy Research: a Comparison of Approaches. *Academy of Management Review*, 11(4), pp. 801–14.

Ward, P. T., Duray, R., Leong, G. K. and Sum, C. C., 1995. Business Environment, Operations Strategy and Performance: an Empirical Study of Singapore Manufacturers. *Journal of Operations Management*, 13, pp. 99–115.

White, R. E., 1986. Generic Business Strategies, Organisational Context and Performance: an Empirical Investigation. *Strategic Management Journal*, 7, pp. 217–31.

Wong, K. and Kwan, C., 2001. An Analysis of the Competitive Strategies of Hotels and Travel Agents in Hong Kong and Singapore. *International Journal of Contemporary Hospitality Management*, 13(6), pp. 293–303.

Wright, P., Kroll, M., Pringle, C. and Johnson, J., 1990. Organisational Types, Conduct, Profitability and Risk in the Semiconductor Industry. *Journal of Management Systems*, 2(2), pp. 33–48.

Yamin, S., Gunasekaran, A. and Mavondo, F. T., 1999. Relationship between Generic Strategies, Competitive Advantage and Organisational Performance: an Empirical Analysis. *Technovation*, 19, pp. 506–18.

Yang, C. and Lu, X., 2000. Holding the Development Pulse of the Times; Looking Dialectically at the Benefits and Disadvantages of Joining WTO – an Interview with Zhang Xuzhi. *International Petroleum Economics*, 8(3), pp. 8–9.

Yeung, W.C., 1999. The Internationalisation of Ethnic Chinese Business Firms from Southeast Asia: Strategies, Processes and Competitive Advantage. *International Journal of Urban and Regional Research*, 23, pp. 103–27.

Newspapers

Kay, J., 1999. Strategy and the Delusion of Grand Designs. Mastering Strategy (1). *Financial Times*, 29 September, pp. 2–3.

Lucas, L., 2000. Made for One Another: the Complex on Jurong Island has now Attracted Investment of more than $21 bn. *Financial Times*, 28 March.

McNulty, S. 2001. Growth Overshadowed by Regional Uncertainties. *Financial Times*, 11 April, p. 1.

Mayer, C., 1999. Corporate Governance is Relevant. Mastering Strategy. *Financial Times*, 11 October.

Mintzberg, H., Ahlstrand, B. and Lampel, J., 1999. Strategy, Blind Men and Elephant. Mastering Strategy (1). *Financial Times*, 29 September, pp. 6–7.

Prahalad, C.K. 1999. Changes in the Competitive Battlefield. Mastering Strategy (2). *Financial Times*, 4 October, pp. 2–3.

Whittington, R., 1999. The 'How' is more Important than the 'Where'. Mastering Strategy. *Financial Times*, 25 October.

Wolf, M., 2003. The World Must Learn to Live with a Wide-Awake China. *Financial Times*, 12 November.

Unpublished sources

Alexander, E. D., 1990. Elements of Strategic Management and Innovation: an Empirical Investigation of the Elements Linking Strategy-Making to Those Elements Needed to Stimulate Innovation within Scottish Firms. Unpublished doctoral dissertation, Glasgow University.

Aosa, E., 1992. An Empirical Investigation of Aspects of Strategy Formulation and Implementation with Large, Private Manufacturing Companies in Kenya. Unpublished doctoral dissertation, University of Strathclyde.

Corrieria, Z., 1996. Scanning the Business Environment for Information: a Grounded Theory Approach. Unpublished doctoral dissertation, University of Sheffield.

Gourlay, D., 1996. Industrial Relations on Offshore Installations. Unpublished doctoral dissertation, The Robert Gordon University.

Hawkins, J. E., 1995. A Strategic Choice Model for Asia-Pacific Shipping. Unpublished doctoral dissertation, University of Plymouth.

Hertz, L., 1980. The Definition of the Small Business, Theory, Mechanics and Practice: a Comparative study of the Laws of the UK, USA, Israel and the People's Republic of China. Unpublished doctoral dissertation, The City University.

Huang, C., 1993. Competitive Advantages, Corporate Strategy and the Internationalization of Chinese State-Owned Manufacturing Enterprises. Unpublished doctoral dissertation, University of Strathclyde.

Quintella, R. G., 1993. The Relationship between Business and Technology Strategies in the Chemical Industry. Unpublished doctoral dissertation, University of Brighton.

Stevens, R., 1994. The Strategic Choices of Small Business and the Canada–United States Free Trade Agreement: an Analysis of the Free Trade Agreement's Effect on the Competitiveness of Canadian Firms in the Apparel, Furniture and Wine Industries and their Strategic Responses in the New Business Environment. Unpublished doctoral dissertation, University of Bradford.

Authoritative sources

Bank of Scotland, 1996. *Oil and Gas Handbook*, 4th edn. BankWest, Edinburgh.

BP Amoco, 2000. *BP Amoco Statistical Review of World Energy 2000*. London.

BP (The British Petroleum Company plc.), 1998. *BP Statistical Review of World Energy.* June 1998. London.
—— 2001. *BP Statistical Review of World Energy 2001.* London.
—— 2002. *BP Statistical Review of World Energy 2001.* London.
—— 2009. *BP Statistical Review of World Energy 2009.* London.
CBBCS (China–Britain Business Council Scotland), 2000. *China Petro/Chem Expo 2000, The Development and Foreign Co-operation of China's Oil and Petrochemical Industry,* Glasgow.
DTI (Department of Trade and Industry), 1996. *Report on the Oil, Gas, and Petrochemical Sectors in South East Asian Markets,* Produced by British Embassies and High Commissions in South East Asia, Compiled by the Commercial Section of the High Commission at Singapore, June 1996. Singapore.
IEA (International Energy Agency), 2000. IEA Examines China's Quest for Worldwide Energy Security. Press Release, 20 March. IEA/Press (OO)3, Paris.
Mackay, T. and Adam, J., 1998. *Prospects for the World Offshore Oil and Gas Industry 1998–2000.* Mackay Consultants Limited, Aberdeen.
PetroMin, PetroMin Magazine, 2000. *Oil and Gas Directory 2000,* 15th edn. Singapore.
Scottish Enterprise (Energy Team), 2002. *The People's Republic of China – a Profile of the Energy Industry,* March, Beijing.
STI (Scottish Trade International), Beijing Office, 1995. *China's Oil and Gas Industry: a Report for Scottish Oil-and Gas-Related Companies,* August, Beijing.
The Concise Oxford Dictionary, 1995. 9th edn, Thompson, D. (ed.). Oxford University Press, New York.
The Economist, 2003. Is China's Economy Overheating? 13 November 2003.
—— 2004. Emerging-Market Indicators. 14 February, p. 114.
The Oxford Dictionary, 1933. *A New English Dictionary on Historical Principles,* Vol. 10. Clarendon Press, Oxford.
The United Kingdom Offshore Operators Association (UKOOA), 1998. *The UK Offshore Oil and Gas Fiscal Regime, the Need for Competitiveness and Stability,* 1–3, June, London.
World Oil's Marine Drilling Rigs '99/2000, 1999. World Oil, Gulf Publishing Co., December, Houston.

Electronic sources

(a) Websites

Baker Hughes, Rig Counts, http://www.bakerhughes.com/investor/rig, 2009.
BAOSTEEL (BAOSTEEL Group Corporation), http://www.baosteel.com/group_e/index.asp, 2009.
Bristow, http://www.bristowgroup.com/, 2009.
Britain Organisation, Malaysia Sector Summaries: Oil, Gas, Refining and Petrochemical. www.britain.org.my/trade/sector_summary/oil&gas.htm, July 2001.

China Ship Industry Association, http://www.Chinaship.cn/, 2009.

CIA (Centre Intelligence Agency), The World Factbook – Singapore. http:// www.cia.gov/cia/publications/factbook/geos/sn.html, 2001.

CNOOC (China National Offshore Oil Corporation), 2005 and 2007 annual reports, http://www.cnoocltd.com/en/Investment_report.aspx?w=bg, 2009.

CNPC (China National Petroluem Corporation). http://www.cnpc.com.cn/ CNPC/gsjs/fzlc/default.htm, 2009.

COHC (CITIC Offshore Helicopter Company Limited), http://www.china-cohc. com/, 2009.

COOEC (China Offshore Oil Engineering Corporation), http://60.28.77.221: 8080/haiyou/english/1gsjj/1gsjj_1gywm.htm, 2009.

COSL (China Oilfield Services Limited), COSL strategy: http://www.cnoocs. com/aboutus/strategy.jsp, 2009.

CSIC (China Shipbuilding Industry Corporation), http://www.csic.com.cn/ en/default.htm, 2009.

CSSC (China State Shipbuilding Corporation), http://www.cssc.net.cn/enlish/ index.php, 2009.

Department of Statistics Malaysia, http://www.statistics.gov.my/eng/index. php?option=com_content&view=article&id=50:population&catid=38: kaystats&Itemid=11, 2009.

EIA (Energy Information Administration). Country Information on Malaysia. http://www.eia.doe.gov/emeu/cabs/malaysia.html, May 2001.

——, Cushing, OK WTI Spot Price FOB (Dollars per Barrel), http://tonto.eia. doe.gov/dnav/pet/hist/LeafHandler.ashx?n=PET&s=RWTC&f=D, 2009.

—— http://tonto.eia.doe.gov/country/country_energy_data.cfm?fips=MY, 2009.

Ensco, http://www.enscous.com/default.aspx, 2009.

Fluor, http://www.fluor.com/Pages/Default.aspx, 2007.

Gulf Coast Oil Directory, http://www.oilonline.com/store/directory.asp., 2002.

Inpex, http://www.ipedex.com.my/, 2007.

MISC (Malaysia International Shipbuilding Corporation), http://www.misc. com.my/, 2009.

National Bureau of Statistics of China, China Statistical Yearbook 2003, http:// www.stats.gov.cn/english/statisticaldata/yearlydata/yarbook2003_e.pdf, 2003

—— Statistical Yearbook, 2007, http://www.stats.gov.cn/english/, 2007.

—— Statistical Yearbook, 2008, http://www.stats.gov.cn/english/, 2009.

Ng Han Wee, Eye on Industry Singapore a Global Hub in Oil and Gas Equipment, http://www.spring.gov.sg/newsarchive/epublications/et/2006_ 01/index4.html, May 2009.

Noble Corporation, http://www.noblecorp.com/, 2009.

Oilonline, http://www.oilonline.com/gcod.html. 2000.

Peopledaily, China Sees Oil and Gas Production Increase, http://english. peopledaily.com.cn/200001/11/eng20000111X109.html, 11 January 2000.

PetroMin, http://www.petromin.safan.com. 1999.

PETRONAS, Annual Report 2008, http://www.petronas.com.my/internet/ corp/centralrep2.nsf/frameset_corp?OpenFrameset, 2009.

PHI, http://www.phihelico.com/, 2009.

Pride (Pride International), http://www.edb.gov.sg/edb/sg/en_uk/index.html, 2009.

Qiao, J. Fight, *Fired between Two Chinese Oil Giants, Economic Information and Agency.* http:// www.tdctrade.com/report/indprof/010204.htm, February 2001.

Schlumberger, Worldwide Rig Activity, http://www.slb.com/rigcounts/index_world.cfm, 2009.

SEACOR, http://www.seabulkinternational.com/aboutus_main.html, 2009.

SeaMetric International AS, Interim report Q4 and full year 2008, www.seametric.com, 25 February 2009.

Sembawang, http://www.sembawangenc.com/, 2009.

Sembcorp, http://www.sembcorp.com/sembcorp/, 2009.

Singapore EDB, http://www.edb.gov.sg/etc/medialib/downloads/industries.Par.33625.File.tmp/Energy%20Factsheet.pdf, 2009.

Sinopec (China Petroleum & Chemical Corporation), http://english.sinopec.com/, 2009.

SPC (Singapore Petroleum Company Limited), http://www.spc.com.sg/home/home.asp, 2009.

Statistic Singapore, http://www.singstat.gov.sg/stats/themes/economy/hist/gdp1.html, 2009.

The Association of Qualitative Research Practitioners (AORP), Qualitative Essentials: When to use qualitative research? http://www.aqrp.co.uk/page1.htm, 6 June 1999.

The Economist – The Economist Intelligence Unit, http://www.economist.com/countries/China/profile.cfm?folder=Profile-Forecast, 2009.

The M-I SWACO Worldwide Rig Count, http://www.miswaco.com/Rig_Count/About_This_Count, 2009.

The UN Statistical Yearbook, http://books.google.co.uk/books?id=qGZaY_Xpm7wC&pg=PA146&lpg=PA146&dq=malaysia+2002-2004+gdp&source=bl&ots=fO2_W5rhhw&sig=LRLJVhgMmhHHfnAJqphd1SKi8Jc&hl=en&ei=sKEdSpGHBouP_Qaw0tjGDQ&sa=X&oi=book_result&ct=result&resnum=4#PPP1,M1

The USA Embassy Singapore, Tyger Burning Bright: Singapore's Oil Industry, http://www.usembassysingapore.org.sg/embassy/politics/pet96.html, May 1997.

The World Bank, WBG (World Bank Group), Seven Percent Economic Growth Predicted for China Next Year, China Gateway. http://www.chinagate.com.cn/pls/ora/english.html, December 2001.

—— http://www.worldbank.org/, 2009.

US Commercial Service, Oil & Gas Oil & Gas–Singapore, Singapore Overview, http://www.buyusa.gov/asianow/soilgas.html, 2009.

Zaobao, http://www.zabao.com/special/realtime/news.html, 26 August 2002.

(b) Email

Croy, G., 2000, Editor of PetroMin Hydrocarbon Asia, Asia Pacific Energy Business Publications Pte Ltd in Singapore. Email: answers to the Interview Guide, 31 August.

—— 2001. Reply: Industry Classification (email) (25 February).

Monographs

BBC Enterprises, 1994. Intelligent Island, TV-Video, London.

Brown & Root Company, 1999. A Powerpoint presentation, the Leading Oil & Gas Conference (LOGIC), 1999, Aberdeen.

Ellix, 2002. A Powerpoint presentation, the LOGIC conference, Aberdeen.

Lee, J., 2000. Practicing Supply Chain Management, Wood Group Engineering (WGE), the LOGIC Conference 2000, Aberdeen.

Phillips, I., 1998, Offshore Management, course handout, Aberdeen.

Simmons & Company International, 1998, 1999. Company Profile, USA.

Tan, Y., 2000. Interview Notes: an Interview with a Chinese Engineering Supervisor, 13 August.

Index

Page numbers in **bold** refer to figures, page numbers in *italic* refer to tables.